APL2 in Depth

Norman D. Thomson
Raymond P. Polivka

APL2 in Depth

Springer-Verlag

New York Berlin Heidelberg London Paris
Tokyo Hong Kong Barcelona Budapest

Norman D. Thomson
Finnock House
Cliff Terrace Road
Wemyss Bay
Scotland PA18 6AP

Raymond P. Polivka
60 Timberline Drive
Poughkeepsie, NY 12603
USA

Library of Congress Cataloging-in-Publication Data
Thomson, Norman (Norman D.)
      APL2 in depth / Norman Thomson, Ray Polivka.
         p.   cm.
      Includes bibliographical references and index.
      ISBN 0-387-94213-0 (softcover : alk. paper)
      1. APL2 (Computer program language)   I. Polivka, Raymond P.
   (Raymond Peter), 1929–   . II. Title.
   QA76.73.A655T56   1995   95-18542
   005.13'3—dc20

Printed on acid-free paper.

Production managed by Laura Carlson; manufacturing supervised by
   Jeffrey Taub.
Camera-ready copy prepared by the authors.
Printed and bound by R.R. Donnelley & Sons, Harrisonburg, VA.
Printed in the United States of America.

9 8 7 6 5 4 3 2 1

ISBN 0-387-94213-0 Springer-Verlag New York Berlin Heidelberg

Preface

This book is designed for people with a working knowledge of APL who would like to increase their fluency in the wide range of extra facilities offered by second-generation APL products. Although the primary product in view is IBM's APL2 as implemented on mainframe, PC and RS/6000, the language features covered share considerable common ground with APL*PLUS II and Dyalog APL. This is a book about skills rather than knowledge, and an acquaintance with some variety of APL on the reader's part is assumed from the start. It is designed to be read as a continuous text, interspersed with exercises designed to give progressively deeper insight into what the authors conceive as the features which have the greatest impact on programming techniques. It would also be suitable as a text-book for a second course in APL2, although experience suggests that most programming language learning is now by self-study, so that this volume is more likely to provide follow-up reading to more elementary texts such as "APL2 at a Glance" by Brown, Pakin and Polivka. Material is discussed more informally than in a language manual - in this book textual bulk is in proportion to difficulty and importance rather than to the extent of technical details. Indeed, some APL2 extensions are not covered at all where the technicalities pose no great problems in understanding and can be readily assimilated from the language manuals.

Second-generation APL is dominated by two ideas - nested arrays and operator extension. Nested arrays are in principle so simple a concept that only a few minutes are needed for an experienced APL user to read and absorb their technical specifications, and also those of the closely associated functions **enclose, disclose** and **depth**, and the operator **each**. Nevertheless the increase in expressiveness and potential complexity which these few simple ideas add is truly astonishing.

The first chapter discusses APL2 arrays and functions, grouping the latter into broad areas such as structuring, selection and inquiry. Chapter 2 considers operators, both primitive and user-defined. Chapter 3 contains demonstrations to show how nested arrays deal with simple data structures in a way which makes their behavior comprehensible and useful to people with very limited programming background and experience. Chapters 4-6 then retrace and develop the ideas of chapters 1-3. Chapter 4 develops the ideas of chapter 1, but focusing more on the way in which functions interact. Chapter 5 develops the Chapter 2 discussion of operators in a similar way, and Chapter 6 gives more sophisticated examples which use all the powerful features of APL2 which have been developed in the previous five chapters.

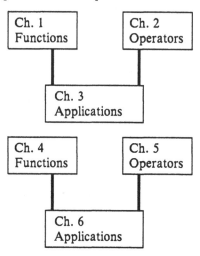

Using APL2 to its full capability is a skill whose acquisition takes time and patience which are an order of magnitude greater than the skills needed for a mastery of first-generation APLs. The reward, on the other hand, is the stimulus of a language whose exploration is a source of constant delight through its seemingly endless capacity for expressing ideas of indefinite complexity in unambiguous and succinct terms.

Of the existing APL texts and primers, some were originally first-generation APL works upgraded to APL2 by the addition of new sections and appendices. Others assume that APL2 is the first language which the user encounters. This book is addressed to the thousands of people, some programmers and some not, who have achieved both enhancement to their professional skills and personal satisfaction in learning and using APL, and who would like to build on this foundation by acquiring a matching fluency in the skills associated with APL2.

Extra weight is therefore given in the text to those APL2 language features which extend user versatility in describing data structures and communicating algorithms in ways which mirror current thinking in computing science and software engineering.

The exercises are designed to give the reader practice in these processes. Frequently, subtler points of difference are best illustrated by exercises with long sequences of similar expressions to be evaluated. With many exercises, a few keystrokes on a terminal will deliver the answer, and whilst the reader is encouraged to use a computer as a check, the fullest value of most exercises is obtained by predicting the result *before* having the computer deliver it. All solutions are given in an appendix so that it is possible to use the book as a study guide even without the availability of an APL2 system.

The functions, operators, and much of the data in the text are available on a 3.5" disk. Either of the authors can supply particulars.

APL2 has its roots in Ken Iverson's original concepts of a symbolic notation for use with computers. Through the efforts over the years of Jim Brown and his team at IBM's Santa Teresa Laboratory these have matured into a language which can be consistently used through the whole range of software development, that is specification, design, coding, and testing. We have attempted to penetrate beyond the mere description of the syntax and semantics of the language and provide a study in greater depth of the interaction between nested arrays and the various functions and operators. We hope that the present work, "APL2 in Depth" will encourage greater use of APL2.

We should like to acknowledge the thorough and thoughtful review by Curtis Jones, without which this text would have been greatly poorer, and also helpful comments and suggestions from Garth Foster, Helmut Engelke, Bert Rosencrantz, Phil Benkard, and Ron Wilks. We are also greatly indebted to Jon McGrew for his invaluable help in the typographical preparation of the text.

Ray Polivka   Norman Thomson
60 Timberline Drive   Finnock House
Poughkeepsie, New York 12603   Cliff Terrace Rd.
USA   Wemyss Bay
                                                Scotland PA18 6AP

Conventions Used for Arrays, Functions, Operators and Indentities

Names of Arrays

A general principle is that "small" arrays, that is those which require fewer than 15 non-blank characters to construct them by direct keyboard entry, are given either meaningful names, or one-character names, usually S for scalar, V for vector, M for matrix, A for array. Larger objects are given either descriptive names or a name such as V23 which denotes the third vector defined in Chapter 2. Objects so named are stored on the disk which is available to accompany the book.

Functions and Operators

When a word like "pick" is used in its specific role as an APL function or operator, it is printed in a heavy font thus : **pick**.

The following conventions are used in writing defined functions and operators:

F	a function
Z	result of a function or derived function
L,R	left and right arguments of a function
P,Q	left and right operands of an operator
T,U,V	local variables

Operations which behave identically, or nearly so, but contain different code are distinguished by using different combinations of upper and lower case letters in their names.

A subsidiary operation is indicated by the prefix Δ.

Labels and Comments

Labels are named L1, L2, L3, ... and so on.
  The general format for a line in an operation (i.e. a function or operator) is:

[line-number]   Ln: expression ⍝ comment

There are several distinguishable uses for comments, in particular they may describe

  (a) constraints on arguments and operands prior to execution;
  (b) description of a result following execution;
  (c) effects on global variables in the workspace;
  (d) clarification in words of a single APL line;
  (e) description of current execution status of data and/or a program.

Comments are permitted on the header line in many APL2 implementations. Where type (a) comments can be expressed sufficiently briefly this usage is adopted, otherwise they are given in separate lines at the head of the function. Most of the functions in this book are very short, and the emphasis is on transforming ideas into APL2 program fragments rather than on the development of programming systems where type (e) comments are more likely to be found. Where these occur in production APL2 code they frequently indicate the possibility of breaking down a function into subfunctions.
  Labels may be *floating*, that is a function line may consist of a label and its colon only, possibly with a comment. Judicious use of the floating label and comment combination can add considerably to legibility, adaptability and maintenance of functions. Floating labels used in this way are another frequent indication of suitable points for subdividing a function into subfunctions.

Identities

The symbol ←→ is used to denote "is identically equal to."

Contents

1
Functions and Arrays in APL2

Compared with first-generation APL, APL2 brought about a vast explosion in the amount of data types and structures which can be modelled. This chapter starts with a discussion of data structures in APL2, beginning with nested arrays and followed by some notes on complex numbers. There then follows a discussion of the principal APL2 primitive functions under the headings of

Construction
Selection
Replacement
Restructuring
Comparison and Enquiry

1.1 Nested Arrays and Depth

An APL2 *array* is an object which possesses two properties namely *data* and *structure*, the latter of which has two measures, namely *shape* and *depth*. An array may be of any dimension, and scalar, vector and matrix are special names describing the special cases of 0, 1 and 2 dimensions respectively. The principal feature which distinguishes second-generation APLs is the concept of nested arrays of which the following is an example:

```
M11←2 2ρ'CHARS' (ι4) (2 2ρ'ABCD') 16
```

A *nested array* is an array in which any item may itself be an array, and at least one item has rank greater than zero. APL2 arrays are distinguished by two characteristics not available in first-generation APL, namely

1. heterogeneity (mixed data types)
2. depth (nested arrays)

DISPLAY is a function which comes in a workspace distributed with the IBM products and which reveals the structure of objects with regard to nestedness, for example:

DISPLAY M11

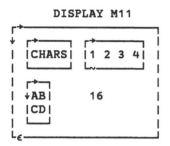

The depth of M11 is given by

$\quad \underline{\equiv}$M11
2

i.e. two is the maximum number of line crossings in DISPLAY M11 required to reach the most deeply nested part of the array. The arrows on a DISPLAY box indicate separate axes, and so the total number of arrows is the rank of the array.

Every array possesses (a) data, and (b) structure, i.e. shape and depth. Shape relates to the data organization at any given level of nesting. Distinction is made between an *item* of an APL2 array, and the *contents* of the item, a term which implies the removal of one level of structure. The contents in general will consist of further APL2 arrays which may themselves possess structure, and so on in a nested fashion. For example the item of M which occupies the first row first column position is a five-item character vector whose contents are the five characters 'C' 'H' 'A' 'R' and 'S'. Evaluation of APL expressions thus involves a *structure* phase which logically precedes evaluation of data values in a *function* phase.

The depth of an array relates to the nesting of the data items. While **shape** (ρ) determines the shape and rank of an object, **depth** ($\equiv$) indicates the degree of nesting within an array. It returns a non-negative integer which defines the *maximum* number of levels of structure to be penetrated in order to get to a simple scalar where *simple* means non-nested.

Here is a more elaborate example:

```
A←'CHARS'
B←ι3
C←'A'2'B'
D←ι0
E←2 2ρ'ABCD'
F←2 2ρ2 4'AB'(ι3)
V11←A B C D E F 5 '5'
DISPLAY V11
```

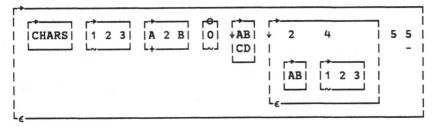

The top left corner of a DISPLAY box contains information about rank and emptiness thus:

→ and ↓ denote the first and subsequent dimensions respectively;
⊖ and Φ denote emptiness, if present, in these dimensions.

In the case of an empty array, DISPLAY exhibits the non-empty dimensions, using the *prototype* of the array to show all items. The prototype of any array is another array which indicates the type and structure of its first item but not its data. If an array has no non-zero dimensions, its DISPLAY box nevertheless indicates its rank (in the case of a scalar by omitting the box altogether), and the contents are either a 0 indicating numeric, or a blank indicating character. The DISPLAY of an empty numeric vector is thus a box containing 0, that of a numeric array with shape vector 0 4 is a box containing a vector of four zeros, and that of an array with shape vector 2 0 4 is a box containing a 2 by 4 matrix of zeros. The function PROTO below defines *prototype*, and is discussed in more detail in Chapter 4. PROTO also illustrates the style of function display which will be followed in this book, that is with no ∇s, and with the header line numbered [0].

```
[0]    Z←PROTO R
[1]    Z←↑0ρR
```

The bottom left corner of a DISPLAY box contains information about type and depth according to the following code:

No symbol   character data
-            scalar blank or character scalar (e.g. final 5 above)
                 when non-scalar arrays are present
~   simple numeric
+   simple mixed character and numeric
ε   nested

1.1.1 Complex Numbers

A further advance with second-generation APLs is the admission of complex numbers. In APL2 complex numbers may be expressed either in Cartesian or in polar form so $(0 + j1)$ can be represented in two equivalent ways:

```
    (0J1)≡1D90
1
```

+C returns the complex conjugate of C. |C returns the magnitude (absolute value) of C.

RA+0J1×IA combines dimensionally compatible arrays RA and IA, representing real and imaginary parts, into a single complex array.

The circle function o is extended to make it easy to carry out standard mathematical operations with complex numbers thus:

left argument			left argument
9	Real part	Imaginary part	11
10	Argument	Phase	12
¯9	C (i.e. null function)	Cj (i.e. C×0J1)	¯11
¯10	conjugate of C (i.e. +C)	exp(Cj)	¯12

If C is thought of as a point in the Argand Diagram with O as the origin, ¯10oC represents the reflection of OC in the real axis, and ¯11oC represents its anticlockwise rotation by one right angle.

9 11•.oC breaks a complex array into real and imaginary parts and its shape is 2,ρC.

9 11oⲤC also breaks a complex array into real and imaginary parts but the result is nested of shape 2.

0J1×+C exchanges real and imaginary parts.

Illustrations : Complex numbers

a. The fourth root of j (0J1) can be obtained in two ways, viz.

```
    (¯1101)*.25
0.92388J0.38268
    0J1*.25
0.92388J0.38268
```

b. The classical equation in mathematics connecting e, π, and j, namely $\exp(\pi j) = -1$, is:

```
    ¯12oo1
¯1
```

c. De Moivre's theorem is illustrated by:

```
      THETA←12o3J1
      (¯12oTHETA)*4        ⍝ (expjθ) to power 4
0.28J0.96
      ¯12o4×THETA          ⍝ exp j4θ
0.28J0.96
```

d. Find the square root of

$$\frac{5-j15}{3-j}$$

and verify the result:

```
      (5J¯15÷3J¯1)*.5
2J¯1
      (5J¯15÷3J¯1)≡2J¯1*2
1
```

1.2 Construction of Arrays

Vectors are no less useful in APL2 than in first-generation APLs. They may be
constructed either

> *explicitly* through a number of functions such as **ravel**, **reshape**, **catenation**
> and **enlist**; or

> *implicitly* through vector notation.

1.2.1 Vector Notation

The standard syntax for constructing numeric vectors from simple scalars is to
separate items with spaces thus:

```
10 20 30
```

It is possible to construct character vectors in the same way:

```
'A' 'P' 'L'
```

as well as in the more common fashion `'APL'`. Vector notation allows any item
to be replaced by a variable name or a parenthesized expression, e.g.

```
A B 20
```

```
10 'A' (X=2)
```

In the last example the parentheses are essential to achieve the required
grouping into three terms. Without them the items form two groups, three of
them in the left argument of = and one in the right.

Such parenthetical groupings may be nested, for example

```
       DISPLAY V12←12 (13 (14 15)) (16 17)
 ┌→───────────────────────────────────────┐
 │                                         │
 │ 12 │  ┌→───────────────┐ │ ┌→─────┐ │   │
 │    │  │      ┌→──────┐  │ │ │16 17│ │   │
 │    │  │ 13 │14 15│  │ │ │ └~─────┘ │   │
 │    │  │      └~──────┘  │ │         │   │
 │    │  │                 │ │         │   │
 │    │  └∈────────────────┘ │         │   │
 └∈───────────────────────────────────────┘

       (ρV12),≡V12
 3  3
```

Parentheses in conjunction with vector notation are used as a form of implicit enclosure. They are non-redundant if they serve both to group and separate, regardless of where they appear in an expression. Vector notation was originally called "strand notation," and the terms are equivalent.

Illustration : Separating and Grouping

Consider the following expressions

```
a.  (10 20)
b.  10 (20)
c.  10 (20) 30
d.  10 20 ((30 40))
e.  10 (⌈5.6) 30
f.  10 (20 30) (40 50)
```

In the first three, the parentheses are redundant, in (a) they group but do not separate, in (b) and (c) they separate but do not group. In (d) one set of parentheses is redundant - the inner ones group but do not separate, while the outer ones separate but do not group. In the non-redundant case (e) the parentheses define a subexpression, while in (f) both sets of brackets both group and separate.

Such distinctions are also indicated in the APL2 default output display by the use of indentation to show depth, for example

```
      10  20  30
10  20  30
      10(20 30)
 10    20 30
```

DISPLAY serves the same function but makes the difference even clearer:

```
       DISPLAY¨(10 20 30)(10(20 30))
```

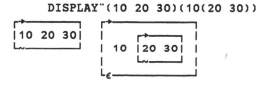

Exercises 1a

1. Sketch the graphic picture which the DISPLAY function would produce for
the following:

a.	`'ABC' 17.6`
b.	`2 3ρ2 2 4`
c.	`2 3 4ρ2 2 4`
d.	`2 4ρ'ABC' '' ' ' '6'(ι2)(ι0)9 6`
e.	`'A' 7.5 5 '5'`
f.	`0 3ρ5`
g.	`0 3ρ'A'`
h.	`0 3ρ 5 'A'`
i.	`3 0ρ 5 'A'`
j.	`0 3ρ(5 'A')4`
k.	`0 3ρ('B'6)(5'A')`
l.	`0 0ρ('B'6)(5'A')`
m.	`0 2 0ρ('B'6)(5'A')`

In each case what is the prototype?

2. Two empty arrays are displayed below which differ in two details. Use the
rules for DISPLAY boxes to find APL2 expressions which could have generated
them.

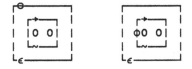

3. Write a monadic function DIS which on the first line displays the shape and
depth of its argument thus

    SHAPE: 2   DEPTH: 3

and on the following lines shows the result of DISPLAYing the argument. This
function can be used to give the total descriptions of APL2 objects which are the
subject of exercises 4-6.

4. a. What are the value, shape and depth of `1(2 3)+(ι3)4` ?

   b. If `A←4 5    B←3    C←'APL'`, what are the value, shape and depth of
`3ρ(A B)C` ?

5. With A and B defined as in qn. 4, what is the difference between

   a. `A B×5 A` and b. `A(B×5)A` ?

6. (i) If `A+2 3ρι6` and `B+3`, all but two of the expressions below are of shape two - which are the two?

a. `(A+1)A`   f.   `(ρA)(ρB)`
b. `A 2`   g.   `((ρA)(ρB))`
c. `A 2 -1`   h.   `A B`
d. `A 2 ¯1`   i.   `'A' 'P' 'L'`
e. `A(2(3 4))`   j.   `'AP' 'L'`

   (ii) All but two are of depth two - which are they?

7. What can be said about the value of `(ρA)1ρA` where `A` is any array? If it has no value, what type of error is generated and why?

8. Within each row which expressions are identical for a general array `B`?

   (i) a. `B+ι3`   b.`(B+ι3)`   c. `((B+ι3))`

   (ii) a. `B(B+1)(B+2)`   b. `B (B+1) (B+2)`   c. `B(B+1)B+2`

   (iii) a. `B Bρ5 6`   b. `B(Bρ5 6)`   c. `(B B)ρ5 6`

9. a. What fact about complex numbers is expressed by the identity

   `(Cx+C) ↔ (|C)*2    ?`

   b. Write a function QUAD whose argument is the vector of coefficients (not necessarily real) of a quadratic equation in descending power order, and whose result is a two-item vector of roots. Use QUAD to display the roots as a two-column matrix with the real parts in the first column and the imaginary parts in the second column. Illustrate by solving $x^2 + x + 1 = 0$.
   Obtain the values of QUAD `1 2J3 4J¯1`. How would you confirm that these were indeed the roots?

1.2.2 Enlist

While the primary means of constructing arrays is the **shape** function, other
primitive functions allow alternative construction techniques, e.g. **enlist** and **ravel
with axis**. Enlist returns a simple (i.e. non-nested) vector whose items are the
simple scalars of its argument in order. It thus removes all nested depth - analogous to the way in which **ravel** reduces dimensionality for simple arrays.

```
      V12←12(13(14 15))(16 17)
      ∈V12
12 13 14 15 16 17
```

1.2.3 Ravel with Axis

In APL2 **ravel** is extended to allow qualification with axes. The qualifier must
be simple and, assuming ⎕IO=1, may be any of

  (a) a positive integer scalar in the range 1 to the rank of the argument;
  (b) a vector of consecutive integers from this range;
  (c) a positive fraction not exceeding one more than the argument rank;
  (d) ι0.

Case (a) means do nothing, that is A≡,[N]A for any valid integer N.

For case (b), ,[ιρρA] is equivalent to **ravel** without axes, so for a three-dimensional array, there are two meaningful cases as illustrated below:

```
      A11
SPARE
A
DIME

NO
THANK
YOU
      ρA11
2 3 5

      ρ⎕←,[1 2]A11
SPARE
A
DIME
NO
THANK
YOU
6 5
      ρ⎕←,[2 3]A11
SPAREA    DIME
NO    THANKYOU
2 15
```

The effect on the shape vector is to merge a consecutive pair of items by multiplication.

Case (c) is similar to **laminate** in that a new axis is inserted whose contribution to the shape vector is 1 (with **laminate** the contribution is 2). In the case of a 3-dimensional array there are four possibilities the first three of which are

```
        DISPLAY,[.1]A11   ⍝ dimension vector = 1 2 3 5
┌┌┌→─────┐
↓↓↓SPARE│
│││A     │
│││DIME  │
│││      │
│││NO    │
│││THANK│
│││YOU   │
└└└──────┘
```

```
        DISPLAY,[1.1]A11   ⍝ dimension vector = 2 1 3 5
┌┌┌→─────┐
↓↓↓SPARE│
│││A     │
│││DIME  │
│││      │
│││      │
│││NO    │
│││THANK│
│││YOU   │
└└└──────┘
```

```
        DISPLAY,[2.1]A11   ⍝ dimension vector = 2 3 1 5
┌┌┌→─────┐
↓↓↓SPARE│
│││      │
│││A     │
│││      │
│││DIME  │
│││      │
│││      │
│││NO    │
│││      │
│││THANK│
│││      │
│││YOU   │
└└└──────┘
```

The fourth possibility of case (c), namely ,[3.1]A11 has the same effect as case (d), that is if the qualifier is ⍳0 a 1 is catenated to the end of the shape vector and the array restructured. In the particular case of vectors the result is a column matrix of shape (⍴V),1. This is a convenient way of converting a row vector into a one-column matrix, e.g.

```
      ,[ι0]εV12←12(13(14 15))(16 17)
12
13
14
15
16
17
```

1.2.4 Default Display of Arrays

The default output routines for mixed character and numeric data use rules that
guarantee a pleasing and intelligible display in the great majority of cases. In
brief, numeric items in columns have decimal points aligned and columns are
right justified unless they contain only character data in which case they are left
justified. The combination of vector notation and these rules makes the writing
of ad hoc reports a great deal easier as the following illustration shows.

Illustration : Writing Reports

```
      ROWS←'FRANCE' 'GERMANY' 'SPAIN'
      COLS←'' 'JAN' 'FEB' 'MAR'
      SALES←3 3ρ52.3 12.95 34 15.3 9.5 12.25 20 35.5 39

      COLS,[1]ROWS,SALES
              JAN    FEB    MAR
 FRANCE   52.3  12.95  34
 GERMANY  15.3   9.5   12.25
 SPAIN    20    35.5   39
```

1.2.5 Enclose and Disclose

Array structure can be created, removed and altered using the functions

> **enclose(⊂)**
> **enclose with axis(⊂[I]),**
> **disclose(⊃),**
> **disclose with axis(⊃[I])**

While vector notation imparts structure to the vector it creates, the **enclose**
function (⊂) is necessary to establish a bounding structural layer around *any*
object other than a simple scalar. The result of enclosure is always a *scalar*. For
example

```
      3 4ρ⊂'APL2'
 APL2   APL2   APL2   APL2
 APL2   APL2   APL2   APL2
 APL2   APL2   APL2   APL2
```

creates a matrix each item of which is the scalar produced by enclosing `'APL2'`.

The vector `V12` of Section 1.2.1 could equally have been created by explicit use of the **enclose** function, viz.

```
V12←12,(⊂13,⊂14 15),⊂16 17

(ρV12)(≡V12)
3  3
```

For simple scalars only it is true that

    `S` is equivalent to `⊂S`

Thus repeated enclosure of a simple (i.e. non-nested) scalar has no effect on it. It is like a cork on water - however hard it is hit, it continues to float. This can be used as a test for simple scalars, and IBM APL2s are sometimes referred to as "floating systems" as opposed to "grounded" systems.

Disclose is the monadic form of `⊃`. It reduces depth throughout an entire object. It removes one layer of nesting (assuming at least one exists) and therefore acts as an inverse to `⊂`:

```
DISPLAY ⊃⊂(1 2)(3 4)
```

Disclose is valid only for arrays whose items at the top level have the same rank, although they do not require to have the same shape. When they do, disclose brings a shape component from the internal structure to the outer structure:

```
        ρ(1 2 3)'APL'
2
        ⊃(1 2 3)'APL'
1 2 3
A P L
        ρ⊃(1 2 3)'APL'
2 3
```

If objects at the topmost level do *not* have the same shape padding is necessary to preserve rectangularity:

```
        ⊃(1 2)3'APL'
1 2 0
3 0 0
A P L

        V12←12,(⊂13,⊂14 15),⊂16 17
        ⊃V12
 12       0
 13  14 15
 16      17
```

Exercises 1b

1. This exercise tests understanding of the floating scalar rule, that is that
S↔⊂S for scalar S.

(i) Are there any differences between the following six phrases when A, B and
C are all numeric scalars?

a. **A,B,C**   d. **((A)(B)(C))**
b. **A B C**   e. **(⊂A)(⊂B)(⊂C)**
c. **(A)(B)(C)**   f. **(⊂A),(⊂B),(⊂C)**

(ii) Repeat the above assuming A, B and C are all two-item vectors, e.g.

   A←1 2   B←10 20   C←3 4

2. If E is **(2 2ρ'X')77(ι5)** what is the difference between

   a. **E,4 5** and b. **E,⊂4 5**   ?

3. If F is **(2 3ρι6)3**, evaluate

a. F   d. 10×F
b. -F   e. 1 2 3×⊂F
c. 1 2+F   f. F×F

4. Create a 2 by 3 matrix which displays as

```
APL2    APL2    APL2
 IS      IS      IS
GREAT   GREAT   GREAT

APL2    APL2    APL2
 IS      IS      IS
GREAT   GREAT   GREAT
```

5. Suppose Z←'' and X←ι3. Describe in detail (that is by giving value, shape
and depth) the values of Z after each step in the following two sequences (a) and
(b) :

   a. **Z←Z,⊂X**   b. **Z←Z X**
       Z←Z,⊂X   **Z←Z X**

Which if any of your four answers are the same ?

6. Distinguish carefully between

   a. **'',⊂'X'** b. **'' 'X'**
   c. **'',' X'** d. **''(⊂'X')**

Which, if any, of these four expressions are identical?

7. If Z←'' and X←ι3 what are

a. `⊃Z,⊂X`   b. `⊃Z X`
c. `⊃Z,⊂'X'`   d. `⊃Z 'X'?`

8. What are the differences between

    a. `,'ABC' 'DE'` and `ε'ABC' 'DE'`
    b. `,(1 3ρ'ABC')'DE'` and `ε(1 3ρ'ABC')'DE'` ?

9. Calendar construction

 a. Write a function **MONTH** which constructs a calendar for a month given the day of the week of day 1 as an integer from 0 to 6 (Sunday is 0,...Saturday is 6) together with the number of days in the month. Head each column with the appropriate three character title `'SUN' 'MON'...'SAT'`. For example

```
       3 MONTH 30
 SUN MON TUE WED THU FRI SAT
                   1   2   3   4
   5   6   7   8   9  10  11
  12  13  14  15  16  17  18
  19  20  21  22  23  24  25
  26  27  28  29  30
```

 b. How would you change the calendar so as to display the weeks vertically?

 c. How would you convert the calendar mixed data type array into one of all characters, e.g. for transmission as an ASCII file?

 d. Assume that a twelve item global vector **DAYS** contains the number of days in each month in a non-leap year. Write a function **START_DAY** which takes as its left argument a Boolean value **LEAP**, as its right argument the day of the week of 1 Jan (as an integer), and returns the twelve item vector **SD** of integers representing the start day of the week for each month.

 e. Use **SD** and **DAYS** to produce the calendar for the entire year. Shape it to appear by quarter, i.e. the first three months appear as the first row, etc.

1.2.5.1 Partial Enclose and Disclose

Formally partial enclose and disclose are known as **enclose** and **disclose with axis**. When an array is enclosed, it becomes a scalar, that is an extra layer of structure is added. Sometimes encapsulating the *whole* array as a scalar is not what is required, but rather enclosure along one or more of its dimensions. For example, given a simple matrix of names, it is useful to be able to create a vector of name vectors. Enclose with axis, ⊂[I]A, accomplishes this, e.g.

```
      M12←3 5ρ'JOHN TED   JASON'
      M12
JOHN
TED
JASON
      ⊂[2]M12
  JOHN   TED     JASON
```

Consider next some examples with numeric arrays:

```
      M←2 3ρι6
      ρM
2 3

      ⊂[1]M                    ⍝ converts matrix into vector of column vectors
  1 4   2 5   3 6
      ρ⊂[1]M
3
      ⊂[2]M                    ⍝ converts matrix into vector of row vectors
  1 2 3   4 5 6
      ρ⊂[2]M
2
```

In the next example the planes of the array **A3** become two nested items:

```
      A←2 3 4ρι24
      ρA
2 3 4
      ρ⊂[3]A
2 3
      ⊂[3]A
  1  2  3  4    5  6  7  8    9 10 11 12
 13 14 15 16   17 18 19 20   21 22 23 24
```

The axis specification is not restricted to a single axis:

```
      ρ⊂[2 3]A
2
      ⊂[2 3]A
  1  2  3  4    13 14 15 16
  5  6  7  8    17 18 19 20
  9 10 11 12    21 22 23 24
```

The rule is that the axes which are specified are those which become nested and thus the following identity holds:

```
      ⊂[⍳⍴⍴A]A  ↔  ⊂A
```

Also the empty vector is acceptable as an axis specification, however it causes
no enclosure:

```
      A  ↔  ⊂[⍳0]A
```

Disclose with axis has the reverse effect to **enclose**. With **disclose** the shape of
the item nested at the top level of the structure become the last dimension of the
disclosed array. Start by defining V as a vector of columns of M :

```
      V←⊂[1]M←2 3⍴⍳6
      V
 1 4   2 5   3 6
      ⍴V
3
      ⊃V                       ⍝ make a vector of vectors into a matrix
1 4
2 5
3 6
      ⍴⊃V
3 2
```

Disclose with Axis, ⊃[I]A, allows the shape of the nested items to be placed in
dimensions other than the last of the newly disclosed array. The axis specified
describes where the nested dimensions will be in the disclosed array. For
example in making a vector of vectors into a matrix the inner-structure shape
vector can be placed either *after* the existing item as in the case above, or *before*
it as in

```
      ⊃[1]V
1 2 3
4 5 6
      ⍴⊃[1]V
2 3
```

The next example shows what happens when the axis qualifier is a vector:

```
      M←3 4⍴⍳12
      MM←M (-M)
      (⍴MM) (≡MM)
 2   2
      ⍴⊃MM
2 3 4
      ⍴⊃[1 2]MM
3 4 2
      ⍴⊃[1 3]MM
3 2 4
```

```
      ⊃[1 3]MM
  1    2    3    4
 ⁻1   ⁻2   ⁻3   ⁻4

  5    6    7    8
 ⁻5   ⁻6   ⁻7   ⁻8

  9   10   11   12
 ⁻9  ⁻10  ⁻11  ⁻12
```

Disclosure is not possible unless all of the items one level down are of the same rank. If however the items do not have the same shape padding takes place as in

```
      ⊃[1](1 2)(3 4 5)
1 3
2 4
0 5
```

The number of integers specified as axes must match the rank of the items, thus the following identity holds:

$$(\rho,I) \equiv \rho\rho\uparrow A$$

In each case the shape vector item or items indexed by the axis goes into the inner structure, and the depth of the result is two.

1.2.5.2 Relationship between Partial Enclosure and Axis Qualifiers

An axis qualifier applied to a scalar dyadic function is equivalent to a combination of enclosure and disclosure along the complementary axes:

```
      M←2 3ρι6
      M
1 2 3
4 5 6
      ⊂[1]M
 1 4   2 5   3 6
      10 20 30+⊂[1]M
 11 14   22 25   33 36
      ⊃[1]10 20 30+⊂[1]M
11 22 33
14 25 36
      10 20 30+[2]M
11 22 33
14 25 36
```

These ideas extend in a natural way to arrays as the following exercise shows.

Exercises 1c

1. a. Defining `M←2 4ρι8` write an expression which transforms `M` into a three-dimensional array each plane of which is a three-times repetition of a row of `M`, i.e.

```
1 2 3 4
1 2 3 4
1 2 3 4

5 6 7 8
5 6 7 8
5 6 7 8
```

b. Defining `V←ι3` write an expression which transforms `V` into a three-dimensional array with two planes and four columns each column of which is `V`, i.e.

```
1 1 1 1
2 2 2 2
3 3 3 3

1 1 1 1
2 2 2 2
3 3 3 3
```

2. If `M13` and `A12` are defined as follows:

```
M13←3 4ρ'ABCDEFGHIJKLM'
A12←2 3 4ρ'ABCDEFGHIJKLMNOPQRSTUVWX'
```

what are value, shape and depth for each of the following :

a.	`⊂M13`	j.	`⊂[1]A12`
b.	`⊃M13`	k.	`⊂[1 2]A12`
c.	`⊃⊂M13`	l.	`⊂[1 3]A12`
d.	`⊂⊃M13`	m.	`⊂[3 1]A12`
e.	`⊂[1]M13`	n.	`⊂[2 3]A12`
f.	`⊂[2]M13`	o.	`⊂[3 2]A12`
g.	`⊂[1 2]M13`	p.	`⊂[ι3]A12`
h.	`⊂[2 1]M13`	q.	`⊂[2 1 3]A12`
i.	`⊂[ι0]M13`		

1.2.6 Partition

Partition is another form of enclosure. **Enclose** (⊂A) forms a scalar of an entire array **A**. **Enclose with axis** (⊂[I]A) forms a set of items by enclosing along specified axes. **Partition** (V⊂A) and **partition with axis** (V⊂[I]A) permit grouping into separate items portions of data along a specific axis where the left argument of partition determines the nature of the enclosure and the axis specification determines the axis along which the partition is to occur.

As an example the following line constructs a three-item vector from a simple numeric vector.

          DISPLAY 1 2 2 2 3 3⊂12 13 14 15 16 17

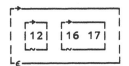

The left argument must be a sequence of non-decreasing non-negative integers, jumps in which correspond to the start of a new partition. In addition zeros may be inserted anywhere to denote that the corresponding items in the right argument are omitted in the result.

          DISPLAY 1 0 0 0 3 3⊂12 13 14 15 16 17

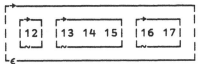

The non-zero items in the left argument do not need to be consecutive, e.g.

          1 1 3 3 7 7⊂'ABCDEF'
     AB CD EF

A partition may have an axis qualifier, so a 5x6 matrix can be made into a 3x6 matrix of vectors by e.g.

          DISPLAY 1 1 3 3 7⊂[1]5 6ρ'ABCDEF'

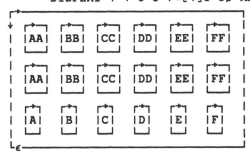

In this case partition with axis reduces the number of rows. Partition along the last axis of a matrix gives a matrix of vectors with a reduced number of columns:

```
        1 1 3 3 7 ⊂[2]6 5ρ'ABCDE'
AB  CD  E
AB  CD  E
AB  CD  E
AB  CD  E
AB  CD  E
AB  CD  E
```

As usual, if no axis qualifier is present, the default is the last axis. In all cases the shape of the left argument must match the dimension along which the partition is to occur.

```
        1 1 3 3 7⊂2 3 5ρ'ABCDE'
AB  CD  E
AB  CD  E
AB  CD  E

AB  CD  E
AB  CD  E
AB  CD  E
```

In the next example, the effect of two successive partitions is to reduce a 2x6x5 array to a 2x4x3 array of vectors:

```
        1 2 2 3 3 4⊂[2]1 1 3 3 7⊂2 6 5ρ'ABCDE'
  AB          CD        E
  AB AB       CD CD     E E
  AB AB       CD CD     E E
  AB          CD        E

  AB          CD        E
  AB AB       CD CD     E E
  AB AB       CD CD     E E
  AB          CD        E
```

Partition applies in the same way to numeric right arguments:

```
        1 2 2⊂[1]3 5ρ⍳15
1       2     3     4     5
6 11    7 12  8 13  9 14  10 15
```

Items in the left argument must be non-negative, otherwise a **DOMAIN ERROR** occurs. As with the vector case above items corresponding to **0** are not carried into the result, e.g.

```
        0 2 2⊂[1]3 5ρ⍳15
6 11    7 12  8 13  9 14  10 15
```

Illustration : Grouping like items

V⊂V on an ordered vector V creates a vector of vectors each containing identical items:

```
V←1 1 1 3 3 5 5 5
DISPLAY V⊂V
```

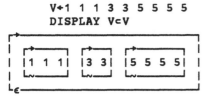

Illustration : Stem and Leaf Plot

Partition provides the foundations of a "stem and leaf" plot, i.e. a pseudo bar chart in which the stems correspond to a range of values and the leaves on each stem are the (possibly rounded) numbers which fall into the corresponding range. Here is an example:

```
      R←?10ρ40
      R
21 34 2 3 22 27 1 16 3 17
      ,[ι0](⌈.1×R)⊂R←R[⍋R]
 1 2 3 3
16 17
21 22 27
34
```

1.3 Selection

The functions **pick, first, index** and **without** provide a variety of means of selecting items from arrays.

1.3.1 Pick and Path

The function **pick** reduces or "penetrates" depth in the sense of going through the levels of structure shown by the DISPLAY of an array.

```
      V12←12(13(14 15))(16 17)
      DISPLAY V12
```

```
r→────────────────────────────────────┐
│                                      │
│ 12 │  r→──────────────────┐  │ r→────┐ │
│    │                      │  │ │16 17│ │
│    │  13 │14 15│          │  │ L~────┘ │
│    │     L~────┘          │  │         │
│    │                      │  │         │
│    │ Lε───────────────────┘  │         │
Lε────────────────────────────────────┘
```

```
      1⊃V12
12
      2⊃V12
13   14  15
```

Pick is like indexing with penetration, that is depth reduction, and so faulty arguments result in **INDEX ERRORS**.

A simple scalar is the only object whose depth is zero.

```
      (≡1⊃V12)(≡2⊃V12)(≡3⊃V12)
0  2  1
```

The left argument of **pick** is a "path" through a nested array. Items in a path should be read from left to right to correspond to penetration of the levels of structure of the object from the outside working in.

```
      2⊃V12
13   14  15
      2 2⊃V12
14  15
      2 2 1⊃V12
14
```

The left argument of **pick** may be a nested vector of depth not more than two, and further the shape of any item in the path must be equal to the rank of the array at that level. For matrices·a nested item in a path is a vector of co-ordinates in an array, e.g. with **M** as defined at the start of the chapter:

```
      DISPLAY M11
```

```
r→──────────────────────────────┐
↓ r→──────┐  r→──────────┐       │
│ │CHARS│  │1  2  3  4│   │
│ L─────┘  L~─────────┘   │
│                         │
│ r→──┐                   │
│ ↓AB │          16       │
│ │CD │                   │
│ L───┘                   │
Lε──────────────────────────────┘
```

the following are legitimate paths:

```
      (2 1)(1 2)⊃M11
B
      (1 2)3⊃M11
3
```

An empty vector in the path is necessary to penetrate a scalar level, e.g. with

        V←'ABC'(⊂ι3)

        DISPLAY V

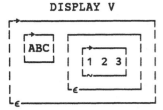

the second item is doubly enclosed so 2 3⊃V is a **RANK ERROR**. In order to
reach the 3 it is necessary to use

        2(ι0)3⊃V
3

An empty vector used in this way can be thought of as a "level-breaker." In
general it is not wise to mix vector notation and explicit enclosure in the same
statement since an extra level of nesting may inadvertently be created.

1.3.2 First

First (↑) is the monadic function which uses the same symbol as **take** but its
semantics are quite different. First penetrates the first level of structure and
produces the first item there, as opposed to indexing which does *not* penetrate.
(This topic is covered in more detail in Chapter 4).

        V12←12(13(14 15))(16 17)

        ↑V12
12
        ↑⌽V12
16 17
        (⌽V12)[1]
  16 17
        DISPLAY (↑⌽V12) ((⌽V12)[1])

        M←2 3ρι6
        MM←M (-M)
        ↑M
1
        ↑MM
1 2 3
4 5 6

If there is no item of **V** within the outermost level of structure, a so-called *fill item* is returned. This is basically a non-empty substitute item for an empty array with zeros where the type is numeric and blanks where the type is character.

```
      DISPLAY ↑0ρ(5 'A')('BCD' 6)
```

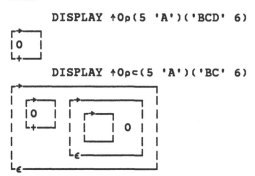

This topic is covered in greater depth in Section 4.3.

1.3.3 Indexing

Two forms of indexing are available in APL2, bracket indexing and scatter indexing. The latter is sometimes informally known as "squad" indexing because of the shape of the symbol (squashed quad). With nested objects the principal difference between **pick** and indexing is that the former reduces depth

```
      V12←12(13(14 15))(16 17)
      2⊃V12
 13  14 15
      ≡2⊃V12
2
```

whereas the latter does not.

```
      2⌷V12
 13  14 15
      ≡2⌷V12
3
```

The quantities **V12[2]** and **2⌷V12** are identical, and both are equivalent to **⊂2⊃V** which suggests that **⊂⊃** and **[]** can be thought of as pre- and post-brackets respectively. In structure terms indexing *cross-sections* arrays whereas **pick** selects items or subarrays from arrays.

For all simple vectors **V** it follows from the floating scalar rule that **V[1]** and **1⊃V** are identical, that is there is no need to distinguish an item and the cell containing it. With nested arrays however this distinction becomes one of crucial importance.

1.3.3.1 Scatter Indexing

This is a versatile facility which nevertheless requires some care in its handling. A basic requirement is that the shape of the left argument is equal to the rank of the right argument as in the following example:

```
      A←3 4ρι12
      A
1  2  3  4
5  6  7  8
9 10 11 12
      3 2⌷A
10
```

The left argument may be nested so that a 2 by 2 cross-section of A can be defined by e.g.

```
      (3 2)(2 3)⌷A
10 11
 6  7
```

The name "scatter indexing" derives from the fact that it is possible to consider the items of an argument such as (3 2)(2 3) as individual indices, and thereby select items one by one using the **each** operator. Although the discussion of operators is the subject of the next chapter, the importance of this case to the **index** function demands its mention here. An example of scatter indexing is

```
      (3 2)(2 3)⌷¨⊂A
10 7
```

Whereas the left argument of **pick** may be indefinitely nested, the depth of the left argument of index may not exceed two. Also it is not possible with either form of indexing to penetrate nested arrays using a single application of the index function, for example:

```
      M11←2 2ρ'CHARS'(ι4)(2 2ρ'ABCD') 16
      DISPLAY M11
```

```
┌→─────────────────────────┐
↓ ┌→─────┐ ┌→───────┐       │
│ │CHARS │ │1 2 3 4 │       │
│ └──────┘ └~───────┘       │
│ ┌→─┐                      │
│ ↓AB│          16          │
│ │CD│                      │
│ └──┘                      │
└ϵ─────────────────────────┘
      1 1⌷M11
CHARS
      ≡1 1⌷M11
2
```

Consider the following attempts to extract the character H from the matrix M11:

```
      2⎕1 1⎕M11
RANK ERROR
      2⎕1 1⎕M11
      ∧∧
      2⎕⊃1 1⎕M11
H
```

These show that repeated applications of indexing alone are not sufficient to extract a nested item. The cause of this apparent dilemma stems from the shape of the result of indexing. Informally the rule is

the shape of the result is the catenation of the shapes of the indices.

Formally for a rank two array M and valid I and J with either form of indexing, if

```
      R1←M[I;J]
      R2←(I J)⎕M
```

then the shapes of the results are

```
      ρR1 ↔ (ρI),(ρJ)
      ρR2 ↔ (ρI),(ρJ)
```

The following phrase selects two copies of the first row of M11, and then the second item within each of these:

```
      ((1 1)2)⎕M11
 1 2 3 4  1 2 3 4
```

With the shape rule in mind, observe the difference between

```
      DISPLAY ((1 1)2)⎕M11
```

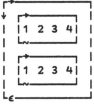

and

```
      DISPLAY ((1 1)(,2))⎕M11
```

The shape of the result of indexing can also be stated in a rank independent fashion. If R←(I J K)⎕A then

```
      ρR ↔ ⊃,/ρ¨I J K
```

Another rule concerning ⎕ is

> *the shape of the index must equal the rank of the array,*

or more exactly, for valid I◻A

$$(\rho,I) \leftrightarrow \rho\rho A\ .$$

Scatter indexing is related to bracket indexing by identities such as the following:

```
    I◻V      ↔    V[I]
  (I J)◻M    ↔    M[I;J]
```

Indexing is highly sensitive to depth and great care must be taken to distinguish situations such as the following:

```
     (((,I),(,J)◻M)  ≡  M[,I;,J]
0
     (((,I) (,J)◻M)  ≡  M[,I;,J]
1
```

1.3.3.2 Indexing with Axes

Axis qualification may be applied to ◻. The axes not included in the axis specification take on all possible values. Thus the second row of M11 in the previous section is

```
      2◻[1]M11
  AB    16
  CD
```

and the second column is

```
      2◻[2]M11
  1 2 3 4   16
```

The first item in the second row can be found in either of two ways:

```
      1◻2◻[1]M11
  AB
  CD
```

or more simply

```
      2 1◻M11
  AB
  CD
```

The latter exemplifies the index rule given in the previous section. Since ◻ identifies *items* and not their *contents* it is not possible to reach the character '**B**' in the matrix by indexing alone. In order to penetrate depth a depth-reducing function such as **disclose** or **pick** must be used, e.g.

```
      1 2[]⊃2 1[]M11
B
      ((2 1)(1 2))⊃[]M11
B
```

The axis qualifier may be a vector of integers corresponding to axes. If M is
extended to three dimensions:

```
      DISPLAY M←M11,[.5]⍉M11
```

the following are valid index expressions:

```
      2 1[][1 3]M      ⍝ 2nd plane, 1st column over all rows
CHARS  1 2 3 4

      2 1[][3 2]M      ⍝ 2nd column, 1st row over all planes
1 2 3 4  AB
         CD
```

The following table of pairs of equivalent expressions should further clarify
the comparison of bracket indexing with [] indexing.

```
                    S←25
                    V←'ABCDEFGH'
                    M←3 4ρ⍳12

          (⍳0)[]S                S
             3[]V                V[3]
       (⊂3 1 2)[]V               V[3 1 2]
           2 1[]M                M[2;1]
     (2 1)(3 4)[]M               M[2 1;3 4]
       (2 1) 3 []M               M[2 1;3]
             1[][1]M             M[1;]
             1[][2]M             M[;1]
        (⊂2 1)[][1]M             M[2 1;]
        (⊂2 1)[][2]M             M[;2 1]
```

1.3.4 Index of

The indexing functions applied to a vector take an index and select the matching data item. **Index of** does the opposite in the sense that it takes a data item and returns the index. If the data item is not found within the vector an integer one greater than the length of the vector is returned, and if the data item appears several times within the vector, the value returned is the index of the first occurrence in the vector.

The left argument of dyadic ι can be thought of as an **alphabet** in which the items of the right argument have to be sought. If the items to be sought are non-scalar care has to be taken to ensure that they are suitably enclosed. The example below illustrates the distinction between seeking the character string `'CHARS'` and seeking the five individual characters `'C'`,`'H'`,`'A'`,`'R'`,`'S'`.

```
      (,M11)ιⲥ'CHARS'
1
      (,M11)ι'CHARS'
5 5 5 5 5
```

Illustration : Character to Numeric Conversion

Character data can be mapped into arbitrary numeric equivalents by defining a left argument for **index of**, thereby providing an elementary coding scheme.

```
      ALP←'ABCDEFGHIJK'
      ALPι'HAD'
8 1 4
      'KJIHGFEDCBA'ι'HAD'
4 11 8
```

Exercises 1d

1. If E is `(2 3ρι6)3'APL'` give value, shape and depth for each of the following (some of the expressions may return errors):

a. `E`	f. `1⊃E`	k. `E[1]`
b. `↑E`	g. `1 2⊃E`	l. `E[1 2]`
c. `↑↑E`	h. `(ⲥ1 2)⊃E`	m. `E[ⲥ1 2]`
d. `ⲥE`	i. `(ⲥ1 2)⊃1⊃E`	n. `(E[1 2])[1]`
e. `⊃E`	j. `⊃ⲥE`	o. `1⎕(ⲥ1 2)⎕E`

2. If W←'ABC' 'DEFG' what are

a. `Wι'ABC'`	e. `Wιⲥ'XY'`	
b. `Wιⲥ'ABC'`	f. `Wι'XY' 'ABC'`	
c. `Wι'DEFG'`	g. `Wιⲥ'XY' 'ABC'`	
d. `Wι'XY'`	h. `Wι(ⲥ'XY')(ⲥ'ABC')` ?	

3. a. With

```
M11←2 2ρ 'CHARS' (ι4) (2 2ρ'ABCD') 16
K←,2
L←,1
```

determine which of the following expressions match `2 1⎕M11`:

1. `(2,1)⎕M11`
2. ` K L ⎕M11`
3. `(K,L)⎕M11`

b. What is the shape of each expression?

4. a. Which of the following expressions does *not* produce a RANK ERROR?

1. `2⎕M11`

2. `1 2⎕M11`

3. `1 1 2⎕M11`

4. `1 (1 2)⎕M11`

5. `(1,1 2)⎕M11`

6. `(1,(1 2))⎕M11`

7. `(1,1,2)⎕M11`

b. Why is a RANK ERROR produced in the remaining cases?

5. With `M←3 4ρι12` determine the value and shape of the following expressions:

   a. `2⎕[1]M`
   b. `2⎕[2]M`
   c. `(⊂2 1)⎕[1]M`
   d. `(⊂2 1)⎕[2]M`

6. Given `A←3 4 5ρι60` write expressions involving ⎕ to

   a. extract the second plane;
   b. extract the third column from each plane;
   c. extract the third column from the second plane;
   d. extract the item in the fourth row, third column, and second plane.

7. a. Use **partition** to divide a sentence, e.g. `'SPARE ME A DIME'`, into a vector of words.

   b. Use **partition** to convert a character name matrix M (i.e. a matrix in which each row is a name, and the shorter names are padded on the right with blanks) into a vector of names, each with no trailing blanks.

   c. Give an expression which converts a vector such as

```
     V←0  0  5  0  0  0  11  0  0  ¯2  0  16  2  0
```

into 0 0 5 5 5 5 11 11 11 ¯2 ¯2 16 2 2 , that is 0 is interpreted as "repeat the last non-zero integer."

8. Write a function ORDINAL which will accept a positive integer and return the character string consisting of the integer followed by the appropriate ordinal representation. For example:

```
     ORDINAL 3
3rd
     ORDINAL 21
21st
```

1.3.5 Without

The function **without** (V~A) provides another way of selecting data from a vector, this time by discarding unwanted items. For example:

```
      'HELLO'~'AEIOU'
HLL
```

Without returns all items of the vector (or scalar) left argument **V** which are not in the right argument **A**. The result is always a vector.

```
      T←'A'~'B'
      T
A
      DISPLAY T
┌→┐
│A│
└─┘

      DISPLAY 'A'~'A'
┌⊖┐
│ │
└─┘

      T≡'A'
0
```

Special attention must be paid to nested arrays since **without** in its comparisons takes into consideration both the shape and structure of items. For example:

```
      K←'A' 'BC'
      DISPLAY K~'A'
```

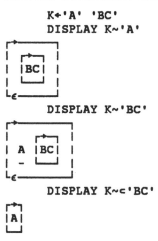

```
      DISPLAY K~'BC'
```

```
┌→──────────┐
│   ┌→─┐    │
│ A │BC│    │
│ - └──┘    │
└∊──────────┘
```

```
      DISPLAY K~⊂'BC'
┌→┐
│A│
└─┘
```

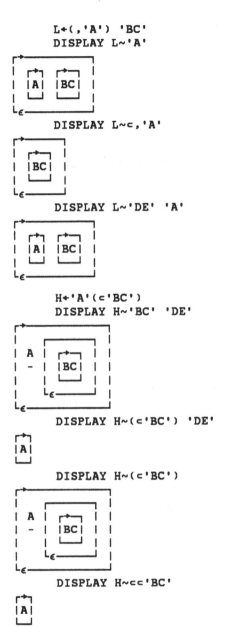

```
      L+(,'A') 'BC'
      DISPLAY L~'A'
```

```
      DISPLAY L~c,'A'
```

```
      DISPLAY L~'DE' 'A'
```

```
      H+'A'(c'BC')
      DISPLAY H~'BC' 'DE'
```

```
      DISPLAY H~(c'BC') 'DE'
```

```
      DISPLAY H~(c'BC')
```

```
      DISPLAY H~cc'BC'
```

In summary, the items of the right argument of the function **without** must reflect the same structure (that is shape and depth) as the left argument if they are to be discarded in the result.

Illustration : Deleting blanks

`V~' '` deletes all blanks in a single string.

Illustration : Intersection of data items

The items common to two vectors are obtained by two successive applications of **without**, e.g.

```
      A←'PICTURE'
      B←'AEIOU'
      A~B
PCTR
      A~A~B
IUE
```

1.4 Replacement

Bracket indexing is the simplest means of replacing parts of APL arrays, but is restrictive in that it is only rectangular subarrays which can be updated. Selective assignment allows much greater generality in updating parts of arrays.

1.4.1 Vector Assignment

Vector assignment allows the decomposition of the assignment target into components each of which can be assigned individually. The general structure of expressions using selective assignment is

    (list of names) ← value(s)

For example:

```
      (A B)←(3 3ρι9)('XYZ')
      A
1    2    3
4    5    6
7    8    9
      B
XYZ
```

1.4.2 Selective Assignment

One form of selective assignment has always been present in APL namely assignment by index:

```
      M←3 3ρι9
      M[2;2 3]←100
      M
1   2   3
4 100 100
7   8   9
```

The ability to assign to just part of an array is greatly extended in APL2. The general structure of expressions using selective assignment is

(selective expression) ← value(s)

The replacement takes place in two steps. The first step selects the items to be replaced and the second does the actual replacement. For example if M is a matrix 1 1⍉M selects the leading diagonal because the left argument of ⍉ asks for matching indices along both dimensions of M.

```
      M←3 3ρι9
      (1 1⍉M)←100
      M
100   2   3
  4 100   6
  7   8 100
```

Illustration : Passing Multiple Arguments

Vector assignment permits the passing of multiple (possibly heterogeneous) arguments as in the opening portion of the following function:

```
[0]   Z← FN NDP;NAME;DEPT;PHONE
[1]   ⍝NDP:   three item vector
[2]   ⍝NAME:  employee name
[3]   ⍝DEPT:  dept name
[4]   ⍝PHONE:phone number
[5]   (NAME DEPT PHONE)←NDP
       •
       •
       •
```

The following is a table of functions allowed in selective assignment:

∈	**enlist**	monadic only
↑	**first/take**	monadic and dyadic with/without axes
↓	**drop**	with/without axes
⌽ ⊖	**reverse/rotate**	monadic and dyadic with/without axes
,	**ravel**	monadic with/without axes
[ ]		bracket indexing as in first-generation APL
⌷		**index** dyadic with/without axes
⍉	**transpose**	monadic and dyadic
⊃	**pick**	dyadic only
ρ	**reshape**	dyadic only
/ ⌿ \ ⍀		derived functions with/without axes

Some derived functions using the operator **each** are also allowed in selective assignment.

The nature of the selective expression can be wide ranging. Suppose V is a vector of vectors nested to an indefinite depth, e.g.

```
V12←(12,(13(14 15))(16 17)
```

and assume that a dyadic function PATH has been written which returns the path in the vector right argument R which leads to the first occurrence of the scalar left argument L, e.g.

```
      14 PATH V12
2 2 1
```

The single item 14 in V12 may have its value changed by selective assignment thus

```
      ((14 PATH V12)⊃V12)←20
      V12
 12    13  20 15    16 17
```

The discussion of how to write PATH as the inverse of **pick** is deferred until Section 5.4.

Illustration : Selective Assignment In Functions

Finding a path to an item in a vector of vectors and simultaneously changing it can be achieved by

```
[0]    Z←L CHANGE R
[1]    ⍝ Find item L[1] in R and change it to L[2]
[2]    ((L[1]PATH R)⊃R)←L[2]
[3]    Z←R

      14 20 CHANGE V12
 12    13  20 15    16 17
```

A variation which enables the two operations of finding and replacement to be performed in the same line is:

```
[0]   Z←L Change R
[1]   Z←↑R((L[1]PATH R)⊃R)←L[2]

      14 20 Change V12
 12    13  20 15    16 17
```

The idea is to return the **first** of R joined to an expression L[2] which is as it were "en passant" assigned to part of R using selective assignment. The effect of right to left execution is that it is the **updated** R which is presented as the argument to **first**. Programmers who find this degree of compression objectionable should nevertheless be able to **recognize** the intention of such code when reading, as opposed to writing, APL2.

1.5 Restructuring

The following functions restructure data:

ρ (**reshape**)	, (**ravel**)	⌽ ⊖ (**rotate/reverse**)
⍉ (**transpose**)	⊂ (**enclose**)	⊃ (**disclose**)
∊ (**enlist**)		

- this section discusses a variety of techniques for doing so.

When constructing a scalar from composite data use of the **enclose** function is probably the technique that comes most readily to mind. This is not however the only way in which data can be reconstructed into scalar form as the following illustration shows:

Illustration : Scalarization

The leading item in an array can be returned as a scalar, possibly enclosed, by applying (ι0)ρ, e.g.

```
      (ι0)ρ(2 3ρι6)('ABCD')
 1 2 3
 4 5 6
```

which has depth two and rank zero. Contrast this with ↑ (**first**) which *selects* the leading item by removing a level of nesting if one exists:

```
      ↑(2 3ρι6)('ABCD')
 1 2 3
 4 5 6
      ρ↑(2 3ρι6)('ABCD')
 2 3
```

⍎(1≥≡A)/'A←⊂A' makes A into a scalar if it is simple and non-scalar, otherwise it does nothing. It should not be used to define the result of a function since if **A** has depth greater than one, the expression reduces to **≢ι0** which although valid does not return a value, so a function applied to it, e.g. ρ≢ι0, gives a **VALUE ERROR**.

There are by contrast situations in which scalars are an embarrassment on account of the floating scalar rule, and it is desirable to eliminate the possibility that an array has empty shape.

Illustration : Descalarization

1/S makes S into a one-item vector if it is a scalar, otherwise does nothing. This can be useful in generalizing algorithms where scalar arguments would result in errors, e.g. routines which use ⊂[ρρA]A to enclose an array **A** along its last axis:

```
      ⊂[ρρA]A←2 2 3 2ρι60
 1  2    3  4    5  6
 7  8    9 10   11 12

13 14   15 16   17 18
19 20   21 22   23 24
```

However a scalar argument results in:

```
      ⊂[ρρS]S←9
AXIS ERROR
      ⊂[ρρS]S←9
      ∧       ∧
```

which can be prevented by:

```
      ⊂[ρρS]S←1/S←9
 9
```

Sometimes it is desirable to increase minimum rank still further as the next illustration shows:

Illustration : Increasing Rank

```
[0]    Z←UPRANK R
[1]    Z←((-2⌈ρρR)↑1 1,ρR)ρR
```

transforms **R** into an array of rank at least two. It is most frequently used as

```
[0]    Z←MATRIFY R
[1]    Z←(¯2↑1 1,ρR)ρR
```

that is make a scalar into a 1x1 matrix, or a vector into a matrix with shape vector 1,ρ**V**.

Yet another restructuring requirement is to create a new array with the shape of
an old one.

Illustration : Copying Structure

A≠A is an all-zeros array of the same structure as **A**.

A∈ι0 is an all-zeros array with the same top-level structure as **A**, e.g.

```
      V12←12 (13 (14 15)) (16 17)

      V12≠V12
0    0  0 0   0 0
      V12∈ι0
0  0  0
```

The **enlist** function may appear in selective specification and so `(∈A1)←A2` uses
the data of **A2** to respecify the values of **A1** whilst still retaining its structure.
   Combinations of functions and operators can be used in selective assignment
as the next illustration shows.

Illustration : Process numerics only in a mixed array

The expression `↑0ρ⊂A` returns the *type* of **A** and is discussed in detail in Section
4.3. It copies the structure of **A** replacing numeric scalars with **0**s and character
scalars with blanks. The following code fragment shows how it can be used in
conjunction with selective assignment and reduction to perform actions on
numeric items only.

```
      A←('XX' 1)('YY' 2)
      (I/∈A)←1.1×(I←∈0=↑0ρ⊂A)/∈A
      A
   XX 1.1    YY 2.2
```

1.5.1 Formatting

The function **format** gives an all character representation, either a vector or a
matrix, of its right argument, and always returns a simple result. Here are some
examples:

```
      DISPLAY ⍕V13←((2 2⍴'ABC')2(3 4)(5 6))
┌→──────────────────────┐
↓ AB   2   3  4   5  6  │
│ CA                    │
└───────────────────────┘
      ⍴⍕V13
2 19

      ⎕←T←⍕J←'ASSETS - ',2.9E6,' EXPENSES - ',2.1E6
ASSETS -  2900000  EXPENSES -  2100000
      (⍴T)(⍴J)
 38   23
```

There are two forms of dyadic format which provide the user with very fine control of the data conversion and display. The first form is called "format by specification" in which the left argument may be either a scalar or vector of integers. If it is a single integer it defines the precision of the character representation of all the numbers in the right argument. If it is a pair of integers, the first defines the total column field width for each column and the second defines the precision for each number. To achieve variation between columns a pair of integers can be provided for each column in the data right argument. For example:

```
      I←2.346 ¯5897.645 .01  0
      J←9 2⍕I
      (⍴I)(⍴J)
 4   36
      I
2.346 ¯5897.645 0.01 0
      J
  2.35  ¯5897.65       .01       .00

      J←6 2 9 1 0 2 5 3⍕I
      (⍴I)(⍴J)
 4   24
      J
  2.35   ¯5897.6 .01 .000
```

The second form of dyadic **format** is "format by example" or as it was originally called "picture format." A simple character vector left argument acts like a template or *picture* describing where the numeric data is to be placed when it is converted to its character representation. This character vector contains both character digits which determine the character representation of the numbers, and also "decorators" which are characters to be displayed in addition to the numbers. The character vector should be viewed as a set of fields, one for each column of the array right argument with each field containing character digits and possibly decorators. The fields are normally separated from each other by at least one blank character.

The decorators may be

> *simple* - that is appearing in the formatted result exactly where placed in the character vector left argument;

controlled - that is the appearance in the formatted result depends upon the number being formatted, e.g. if it is positive or negative; or

floating - position is controlled by the character digits in the associated field.

Each of the ten digits, '0' through '9', has a distinct meaning in the left argument. Full descriptions are given in the language reference manuals, however the following summary may be helpful:

The digits '0','5','8' and '9' form a group.

'0' means display all digits including zeros.
'5' means remove leading/trailing zeros.
'8' means pad with the default format control character (□FC[3]).
'9' means pad with blanks.

The following examples illustrate how these digits can be used to control the display:

```
        C←23.758 0 8653.2

        ' 0000.00'⍕C
0023.76 0000.00 8653.20
        ' 5555.55'⍕C
  23.76         8653.2
        ' 5550.55'⍕C
  23.76    0    8653.2
        ' 5550.00'⍕C
  23.76    0.00 8653.20

        ' 8880.00'⍕C
**23.76 ***0.00 8653.20
        ' 9990.00'⍕C
0023.76    0.00 8653.20
        ' 9990.55'⍕C
0023.76    0    8653.2
```

Next here is an example of a simple decorator |

```
        '| 0000.00'⍕C
| 0023.76| 0000.00| 8653.20
```

The digits '1','2','3' and '4' form a group which handle controlled and floating decorators. Decorators may appear either to the left or to the right of their number or to **both** left and right, in which case two digits from this group should be used which are interpreted left then right. The meanings of the digits are:

'1' : apply a floating decorator to negative numbers only.
'2' : apply a floating decorator to positive numbers only.
'3' : apply a floating decorator to positive and negative numbers.
'4' : cancel '1', '2' or '3' on the other side of the decimal.

Here are some examples to highlight these differences:

```
        D←¯23.758 0 8653.2
        ' $5,551.50CR'⍕D
     ¢23.76CR          .00      8,653.20
        ' $5,552.50CR'⍕D
      23.76         $.00CR $8,653.20CR
        ' $5,531.50CR'⍕D
     ¢23.76CR          $.00      $8,653.20
        ' $5,531.40CR'⍕D
     $23.76CR        $.00CR $8,653.20CR
        ' ¢5,514.50CR'⍕D
     $23.76CR          .00CR  8,653.20CR
```

The digits '6' and '7' deal with the special cases of display of dates and times:

```
        '0006/06/06 06:06'⍕2000 01 01 12 30
     2000/01/01 12:30
        '0006/06/06 06:06'⍕2000 01 01 12 30
     2000/01/01 12:30
```

.. and of numbers in scientific notation:

```
        E←1753.4 ¯.0024 0 ¯284
        '   -1.70↑¯01'⍕E
      1.75↑ 03  -2.40↑¯03    0.00↑ 00  -2.84↑ 02
```

The digit '6' marks the end of a field which is terminated by the immediately following decorator. The symbols , . - are not allowed as decorators in this context.

In the example with digit '7' the symbol ↑ has been used in place of the conventional symbol E.

1.5.1.2 Default rules for mixed data type

The display of arrays of mixed data type is related to the result of the monadic form of format. There is a default pattern whereby the alignment of each column is independent of the contents of other columns, viz.:

1. If the column contains all character data, the column display is left aligned.

2. If the column contains all numeric data, the column is aligned on the decimal point and the decimal digits are right aligned.

3. If the column contains character data and numeric data, the column display is right aligned.

4. If the column contains a complex number, the character data is right aligned with the imaginary value and the J symbols line up. (Complex numbers with D and R symbols are converted to J type numbers on display.) A blank is provided for strictly real numbers. Both the real and imaginary parts align on the decimal point.

WW as defined below ...

```
      W1←'ABC' 'DEF' 'GHI'
      W2←2345 233345 .2345
      W3←1.234 27 98765.43
      WW←3 3ρW1,W2,W3
      WW
 ABC       DEF        GHI
2345       233345     0.2345
   1.234       27 98765.43
```

... appears identical in its output display to ⍕WW:

```
      ⍕WW
 ABC       ABC        ABC
2345       233345     0.2345
   1.234       27 98765.43
```

However:

```
      WW≡⍕WW
0
```

The difference is that ≡WW is 2 while ≡⍕WW is 1. Since the result of format is always simple ⍕A provides a guaranteed means of denesting arrays.

Illustration : Convert an array of arbitrary rank into a matrix

0 1↓0 ‾1↓⍕⊂UPRANK A (see Section 1.5 for UPRANK) transforms any array A into a simple character matrix.

1.5.2 Sorting

Grade-up and **grade-down** may take a left argument provided that the right argument is a simple non-scalar character array. In this case the left argument defines an alphabet or *collating sequence*. Where the collating sequence is a simple vector, it defines "alphabetical order" in the normal usage of that term. For example:

```
      M14←5 3ρ'COWBEEYAKCATSOW'
      M14
COW
BEE
YAK
CAT
SOW
      'ABCEKOSTWY'⍋M14
2 4 1 5 3
      M14['ABCEKOSTWY'⍋M14;]
BEE
CAT
COW
SOW
YAK
```

The collating sequence may have rank greater than one, in which case it is the *last* axis which is the most significant. This means that if a rank two collating matrix is supplied as left argument, all characters in its first column precede any in the second column, all of which precede any in the third column and so on. Suppose a collating matrix COLSEQ is defined as

```
      □←COLSEQ←3 2ρ'BAYSOC'
BA
YS
OC
```

and used to order the rows of M14. Look first at the initial characters of each row of M14. Since C and S are absent from the first column of COLSEQ, CAT, COW and SOW all follow BEE and YAK. The priority of BEE and YAK is judged by observing their second characters. A and E are both absent from the first column of COLSEQ, however A is present in the second column from which E is absent and so YAK precedes BEE. To order the other three rows observe that S precedes C in the second column of COLSEQ and so SOW precedes both COW and CAT whose order is determined by the first column of COLSEQ in which O is present but A is not.

```
      M14[COLSEQ⍋M14;]
YAK
BEE
SOW
COW
CAT
```

The above sounds complicated but the rationale is made clear by considering the most commonly used rank two collating matrix which is

```
abcdefghijklmnopqrstuvwxyz
ABCDEFGHIJKLMNOPQRSTUVWXYZ
```

This matrix represents two collating sequences, the major one along the last axis and the secondary one along the first axis so that all names beginning with the same latter are grouped together regardless of case.

Illustrations : Alphabetic sorting of vectors and matrices

The *Atomic Vector* is a system variable which contains the 256 EBCDIC code representations of the APL2 character set. Letters of the alphabet in the same case occur in natural order in ⎕AV which leads to the following technique for sorting either vectors of words or matrices whose rows are words:

```
      V14←'SPARE' 'ME 'A' 'DIME'
      V14[⎕AV⍋V14]
 A DIME ME SPARE

[0]   Z←SORTC R
[1]   Z←R[⎕AV⍋R;]

      SORTC⊃V14
A
DIME
ME
SPARE
```

For a three-dimensional character array dyadic **grade-up** sorts the array by planes:

```
      ⊃(⊃V14)(M15←3 4ρ'NOT A   CENT')
SPARE
ME
A
DIME

NOT
A
CENT

      ρT←⊃(⊃V14)M15
2 4 5
      T[⎕AV⍋T;;]
NOT
A
CENT


SPARE
ME
A
DIME
```

A related system variable ⎕AF returns either the ⎕AV character given the index or the index given the character, that is it is either ⎕AV⍳R or ⎕AV[R].

There is a **Default Collating Sequence** DCS in the form of a rank three array which is provided in a workspace UTILITY distributed with IBM APL2s. This has the property that the letters of the alphabet occur in alternate case order, i.e. AaBb ... etc.. Its shape is 10 2 28 and its major diagonal plane is:

```
      DISPLAY 1 1 2⍉DCS
```
```
┌→─────────────────────────┐
↓ ABCDEFGHIJKLMNOPQRSTUVWXYZ0│
│ abcdefghijklmnopqrstuvwxyz │
└──────────────────────────┘
```

If DCS is used as the left argument of a **grade** function the right argument must be an array of rank two or above. Using DCS for ordering character data has the advantage that the same letters in different cases are grouped together as opposed to ⎕AV ordering in which all the letters in one case precede any of the letters in another. Also numeric characters appear in numerical order rather than in character order as happens using ⎕AV, that is '9' precedes '10' using DCS.

When discussing numeric vectors, it avoids ambiguity to use the word "ranking" rather than "rank" to denote positions of items following either ascending or descending ordering. These rankings are given by ⍋⍋ and ⍒⍒ respectively:

```
      ⍋⍋12 67 43 28 9
2 5 4 3 1
      ⍒⍒12 67 43 28 9
4 1 2 3 5
```

When there are equal values in a vector ⍋⍋ and ⍒⍒ process the items in order of appearance from left to right within it:

```
      V←5 3 3 5 2 5 9
      ⍋⍋V
4 2 3 5 1 6 7
      ⍒⍒V
2 5 6 3 7 4 1
```

This may not always be the desirable thing to do, so two alternative techniques are shown in the illustrations below:

Illustrations : Averaging tied rankings

```
[0]    Z←TUP R
[1]    Z←.5×(⍋⍋R)+⍒⍋⌽R

[0]    Z←TDOWN R
[1]    Z←.5×(⍒⍒R)+⍒⍒⌽R

       TUP V←5 3 3 5 2 5 9
5 2.5 2.5 5 1 5 7
       TDOWN V
3 5.5 5.5 3 7 3 1
```

Schoolmaster's Rank

Each group of "students" with equal scores is given the highest rank available.

```
[0]    Z←SCH R
[1]    Z←(⊂R⍳R)⌷⍋⍒R

       SCH V
2 5 5 2 7 2 1
```

The combination of **partition** and **grade** allows a simple vector to be reorganized as a vector of vectors where the items within each vector correspond to the same integer in a grouping vector, and zero represents omission:

```
       GV←1 2 1 2 0 3 1 0 2
       R←'ABCDEFGHI'
       GV[⍋GV]⊂R[⍋GV]
 ACG BDI F
```

1.6 Comparison and Inquiry

The functions **depth** and **find** provide means for inquiry of nested arrays, while the **match** function provides a mechanism for their comparison.

1.6.1 Depth

Depth, the monadic function associated with the ≡ symbol, has already been encountered informally as the number of line crossings in the DISPLAY diagram required to reach the deepest part of an array. More formally the depth of an array is defined recursively as one more than the depth of its deepest item, and the depth of a scalar is zero. A simple array is defined as an array with the property that all its items are scalars, and hence it follows that the depth of a simple array is one.

1.6.2 Match

Match is the dyadic function associated with ≡ symbol. Its result is always a simple Boolean scalar which is 1 only if its arguments are equal in all respects, i.e. value, type, shape and depth. Here are some examples of similar objects which fail to **match**:

	arguments differ only in ...
`    2 3≡3 2` `0`	value
`    ''≡ι0` `0`	type
`    (1 1ρ4)≡1 1 1ρ4` `0`	rank
`    (,/1 2)≡⊂⊂1 2` `0`	depth
`    (,/1 2)≡1 2` `0`	rank and depth

and here is one which *does* **match**:

```
    (,/1 2)≡⊂1 2
1
```

The **match** function differs from the scalar function **equal** which *pervades* structure and does comparisons on simple scalar items. The **match** function by contrast does a total comparison on all the attributes of its arguments, thus:

```
    (2 3)4=(2 3) 4
 1 1  1
    (2 3)4≡(2 3) 4
1
```

```
    V←'THE' 'CAT'
    V[2]=⊂'CAT'
1 1 1
    V[2]≡⊂'CAT'
1
```

Match can thus often be used to shorten comparisons as in the following illustration :

Illustration : Test for all items in a vector the same

```
    V≡1⌽V
```

is an alternative to

```
    ∧/V=1⌽V   or   V∧.=1⌽V
```

1.6.3 Find

The **membership** function ∊ tests whether a set of items is contained within another set by reporting the presence or absence of each item of the left argument in the right, e.g.

```
    2 4 7∊ι6
1 1 0
```

To determine if an *entire array* is present in another array requires the dyadic function **find** (⊆) which looks for occurrences of the entire array left argument in the array right argument. The result is a binary array whose shape is that of the right argument with 1s indicating the beginning of occurrences of the left argument.

Illustration : Find all occurrences of one string within another

```
      'CAT'⊆'BATTY CATS SCATTER DUCATS'
0 0 0 0 0 0 1 0 0 0 0 0 1 0 0 0 0 0 0 0 0 1 0 0 0
```

Illustration : Delete Multiple Blanks

```
      V15
 NO    ONE IS    AT   HOME
     (~'  '⊆V15)/V15
NO ONE IS AT HOME
```

The next illustration shows how a matrix can be used as a left argument of ⊆:

Illustration : Pattern Matching

Search for the 2 x 2 identity matrix in a pattern of bits:

```
      M16
0 1 0 1 0 0 1 1
1 0 1 1 0 0 1 1
0 0 0 0 1 1 1 1
1 0 1 0 1 1 1 0
0 1 0 1 1 1 0 0
1 1 1 1 0 1 1 0
0 1 0 0 0 0 0 1
1 1 0 1 1 1 0 1
      PAT
1 0
0 1

      PAT⊆M16
0 1 0 0 0 0 0 0
0 0 0 1 0 0 0 0
0 0 0 0 0 0 0 0
1 0 1 0 0 0 0 0
0 0 0 0 0 0 0 0
0 0 0 0 0 0 1 0
0 0 0 0 0 0 0 0
0 0 0 0 0 0 0 0
```

Exercises 1e

1. Are the following scalars simple? If not what is their depth?

    a. `(ιO)ρ(3 4 5)`   b.   `(ιO)ρ(3 4 5)(6 7)`

2. a. Write an expression to delete leading, trailing, and multiple blanks simultaneously from a simple character vector.

   b. Write an expression using **without** to remove any all-blank rows from a (possibly nested) matrix M.

3. Given the following collating sequences:

```
CS1←' ABCDEFabcdef'
CS2←' AaBbCcDdEeFf'
CS3←⊃' ABCDEF' 'abcdef'
```

and the following matrix:

```
        M17
CAB
DAD
BED
Bed
bed
BAD
ACE
bad
ace
dad
```

determine the results of the following expressions:

    a. `M17[CS1▲M17;]` b. `M17[CS2▲M17;]` c. `M17[CS3▲M17;]`

4. a. Use dyadic **grade-up** to write a function which puts the rows of a character matrix in alphabetical order. Distinguish two cases:

   (i) all upper case letters come before any lower case letter;
   (ii) all a's in any case come before any b's and so on.

   b. Extend your expression to remove duplicate rows.

5. Predict the result of `' '⊆C` where C is a character matrix.

6. a. Generate the matrix of shape R whose top leftmost submatrix of shape L consists of 1s and the remainder is 0s.

   b. Generalize this to the case where the window starts in row W and column C.

7. How would you find where dense points are located in a three dimensional bit array **A** where "dense" means there is a two by two by two cube of all 1's, e.g. if **A13** is

```
0 1 0 1 0
0 1 1 1 0
1 1 1 0 1
1 1 0 0 0
1 1 1 1 1

0 1 0 1 1
1 1 1 1 0
1 1 1 0 0
1 1 1 1 0
1 1 0 0 1
```

then the first plane of the result is

```
0 0 0 0 0
0 1 0 0 0
1 0 0 0 0
1 0 0 0 0
0 0 0 0 0
```

8. a. Write a function **REPL** which replaces each blank in a character array with the character '*'. Test that your function works with arguments of any rank (including 0).

b. What changes would you make to create the function **Repl** which replaces every 0 in a simple (i.e. non-nested) numeric array with the three characters **'N/A'** ?

Summary of Functions used in Chapter 1

Section 1.1
PROTO   prototype

Exercises 1a
DIS   enhanced form of DISPLAY
QUAD   solution of quadratic equations

Exercises 1b
MONTH   calendar for month
START_DAY   auxiliary function for calendar

Exercises 1d
ORDINAL   constructs ordinal numbers

Section 1.4.2
CHANGE   exchanges items in nested vector

Section 1.5
UPRANK   increases rank of array to at least two
MATRIFY   makes scalar or vector into matrix

Section 1.5.2
SORTC   sorts character matrix
TUP   tied upward rank
TDOWN   tied downward rank
SCH   schoolmaster's rank

Exercises 1e
REPL   replaces items in character array

2
Operators

All programming languages contain a set of fundamental instructions to transform data which are collectively referred to as *operations*. In APL2 operations are subdivided into two categories, functions and operators. Data transformation occurs directly through functions or indirectly through operators. The following figure illustrates the relationship of functions and operators.

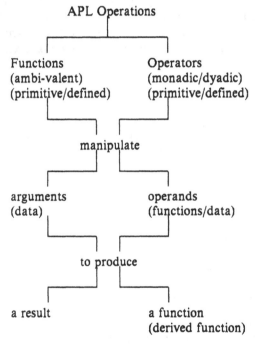

The role of operators is to modify functions before they are applied to data. Primitive operators are discussed in this chapter, and user-defined operators are introduced in Chapter 5. There are four symbols which are used to construct the primitive operators in APL2, namely / \ . ¨ , and there are a total of eight essentially distinct operators which are given in the table below in which P and Q represent functions and V represents a scalar or vector.

P/...	Reduce
V/...	Replicate
P\...	Scan
V\...	Expand
...•.P...	Outer Product
..P.Q...	Inner Product
P¨ ...	Monadic Each
...P¨...	Dyadic Each

2.1 The Each Operator

The **each** operator is intimately connected with nested arrays. It allows functions to be applied item by item to their arguments which is what happens in any case with scalar functions, i.e. functions like +, ÷ and ⊛. **Each** allows its function operands to be applied one level down in the structure of the array. Just as a function acts on data to produce further data called a *result*, so an operator acts on *operands* to produce a function called a *derived function*.

2.1.1 Pervasiveness

A *pervasive* function is one which penetrates the structure of its arguments and is applied to the simple scalars within it. All the scalar functions are pervasive, e.g.

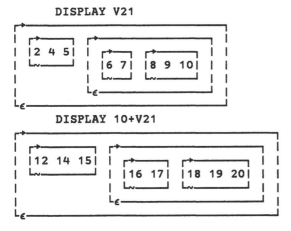

The **each** operator causes its function operand to penetrate one level of structure, and so all **each**-derived functions are pervasive through one level. Repeated applications of **each** are necessary to penetrate further levels of structure:

```
      V22←'ABCD' ('EFG' 'HIJKL')
      DISPLAY 3ρV22
```

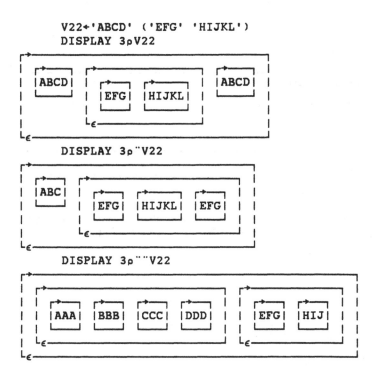

```
      DISPLAY 3ρ¨V22
```

```
      DISPLAY 3ρ¨¨V22
```

There is a formal analogy between the identity

$$S \leftrightarrow \subset S$$

for simple scalars and the identity

$$(F\ R) \leftrightarrow F¨R$$

which defines a pervasive function F, in the sense that after removing R and the parentheses there is a one-to-one correspondence between the symbol sets (S,⊂) on the one hand and (F,¨) on the other.

The **each** operator takes on its real significance when applied to non-pervasive functions such as ι and ⌽ as the following examples show:

```
      ι¨1 2 3 4
 1   1 2   1 2 3   1 2 3 4

      ⌽'APL' 'IS' 'GREAT'
 GREAT IS APL
      ⌽¨'APL' 'IS' 'GREAT'
 LPA SI TAERG

      V12←12(13(14 15))(16 17)
      3 2⌷¨⊂V12
 16 17      13   14 15
```

Each is permitted in selective assignments, e.g.

```
      (↑¨V12)÷2 3 6
      V12
2    3  14 15   6 17
```

Illustration : Multi-path selection (scatter picking)

```
      V12+12(13(14 15))(16 17)
      (2 2 1)(3 2)⊃¨⊂V12
14 17
```

Illustration : Frequency Distributions

Section 1.2.6 illustrated how to use **partition** to group like items. This technique can be developed to obtain a frequency distribution of a vector of integers:

```
      V+2 7 4 5 7 5 4 3 7 7 2
      ρ¨V⊂V+V[▲V]
2    1  2  2  4
```

Illustration : Mid-points in Euclidean geometry

Suppose

```
      (A B C D)+(0 0)(1 6)(5 4)(8 0)
```

represent the co-ordinates of four Euclidean points and the function MIDPT is

```
[0]     Z+MIDPT R
[1]     Z+.5×+/R
```

The mid-points of the sides of the quadrilateral ABCD are:

```
      MIDPT¨(A B)(B C)(C D)(D A)
```

2.1.2 Scalar Extension

Scalar extension applies to the primitive scalar functions which means that if one of the arguments of a scalar dyadic function is a scalar, it is used as many times as necessary in order to apply the function once for each item in the other array argument. Thus the scalar 10 in the expression 10×ι5 is replicated five times to achieve the item by item multiplication. By enclosing one argument scalar extension of non-simple arguments becomes possible, e.g.

```
      V21←(2 4 5)((6 7)(8 9 10))
      DISPLAY (⊂1 3 5)+2 4
```

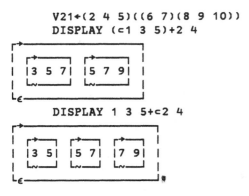

```
      DISPLAY 1 3 5+⊂2 4
```

These expressions can be represented by the following diagrams in which each rectangle or square represents a scalar:

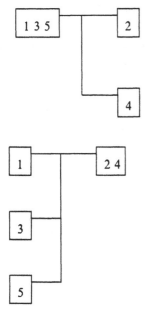

Pervasiveness of scalar functions means that **each** is not required to qualify + in the above expressions.

Derived functions resulting from reduction are *not* pervasive, e.g.

```
      +/(1 3)(2 4)
```

is equivalent to 1 3 + 2 4, that is 3 7, whereas

```
      +/¨(1 3)(2 4)
```

is the result of applying +/ to each of 1 3 and 2 4, namely 4 6. In general if F is a scalar function the following are true:

```
      F/V   ↔  V[1] F V[2] F ...

      F/¨V  ↔  (F/V[1])(F/V[2]) ...
```

Exercises 2a

1. Given

```
V23←('ABC')((ι3)(2 3ρ'ABCD'))
V24←(⊂'ABC')((ι3)(2 3ρ'ABCD'))
```

what are the DISPLAYed values of:

a.　ρV23　　　　　　　　　d.　ρV24
b.　ρ¨V23　　　　　　　　e.　ρ¨V24
c.　ρ¨¨V23　　　　　　　f.　ρ¨¨V24　　　　　•

2. Evaluate the following:

a.　+/(3 4 5) 6 7 8　　　　　d.　+/¨(3 4 5) 6 7 8
b.　+/(3 4 5)(6 7 8)　　　　e.　+/¨(3 4 5)(6 7 8)
c.　+/¨3 4 5 6 7 8

3. Given that V is a lower case name vector, e.g. 'dick' 'anne', replace the first item in each name with its corresponding capital letter?

4. The expression (N+/V)÷N gives the N-period moving average of a vector V. Adapt this to obtain the weighted moving average of V given a vector of weights W which sum to 1, e.g. given weights .2 .4 .4, the weighted moving average of 2 8 5 6 3 1 is 5.6 6.0 4.6 2.8.

5. Given the vector

```
V25←'ABC' (ι3)(ι5) ('THIS' 'IS' 'A' 'TEST')
```

identify the following rearrangements of V25 from the options given below.

```
          V1
ABC
1 2 3
1 2 3 4 5
  THIS IS A TEST


          V2
A    1    1    THIS
B    2    2    IS
C    3    3    A
          4    TEST
          5
```

```
            V3
A B C     1  2  3    1  2  3  4  5    T  I  A  T
                                     H  S     E
                                     I        S
                                     S        T

            V4
A 1 1    THIS
B 2 2    IS
C 3 3    A
  0 4    TEST
  0 5

            V5
ABC   1 2 3   1 2 3 4 5   THIS
                          IS
                          A
                          TEST

            V6
A    1   1      THIS
B    2   2      IS
C    3   3      A
         4      TEST
         5
```

1. V1 ← _____   a.   ⊃[1]V25

2. V2 ← _____   b.   ,[⍳0]¨V25

3. V3 ← _____   c.   ,[⍳0]¨¨⊂V25

4. V4 ← _____   d.   ,[⍳0]¨¨V25

5. V5 ← _____   e.   ⊃[1]¨V25

6. V6 ← _____   f.   ,[⍳0]V25

                          g.   ⊃¨V25

2.1.3 Each with non-pervasive Functions

The effect of applying **each** to the function **shape** is discussed in some detail, following which its application to **index of** and **grade-up** are described in illustrations.

  With 2 5 as left argument and 3 4 as right argument the result of 2 5ρ3 4 is

```
3 4 3 4 3
4 3 4 3 4
```

APL2 however allows us to "scalarize" either or both arguments by enclosure, thereby increasing the possible interpretations of "reshape" for given left and right arguments by using the derived function ρ¨. Scalarizing 2 5 can be pictured structurally as

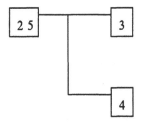

i.e. (2 5ρ3)(2 5ρ4) which is given in APL2 by

```
     (⊂2 5)ρ¨3 4
3 3 3 3 3    4 4 4 4 4
3 3 3 3 3    4 4 4 4 4
```

Enclosure is thus a device allowing scalar extension in the first-generation APL fashion, viz:

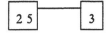

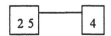

and is therefore an appropriate way to solve the programming problem "Construct two 2 x 5 matrices one made up of 3s and the other of 4s."
  Scalarizing the 3 4 is pictured analogously as

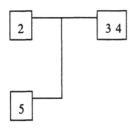

i.e. (2ρ3 4) (5ρ3 4) and is rendered by

```
      2 5ρ¨⊂3 4
3 4   3 4 3 4 3
```

Next the items may be applied pairwise:

which is equivalent to (2ρ3) (5ρ4) and is given by

```
      2 5ρ¨3 4
3 3   4 4 4 4 4
```

Illustration : Each with index of

Obtaining the letter indices of two words using the same alphabet means that the alphabet must be scalarized:

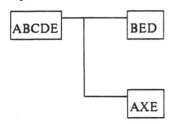

so enclose left:

```
      ( ⊂'ABCDE' ) ι ¨ 'BED'  'AXE'
   2 5 4   1 6 5
```

To obtain the letter indices of a word using two different alphabets it is the *word* which must be scalarized, so enclose right:

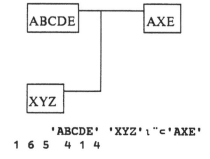

```
      'ABCDE' 'XYZ'ι¨⊂'AXE'
1 6 5   4 1 4
```

Illustration : Each with grade

Two examples are given to deal with multiple words and multiple alphabets.

One alphabet, two words:

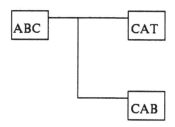

```
    (⊂'ABC')⍋¨'CAT' 'CAB'
2 1 3   2 3 1
```

One word, two alphabets:

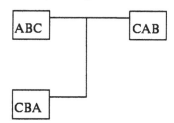

```
    'ABC' 'CBA'⍋¨⊂'CAB'
2 3 1   1 3 2
```

Exercises 2b

1. If `W←'ABC' 'DEFG'` what are

a. `2 3ρ¨W`   d. `(⊂2 3)ρ¨W`
b. `2 3ρ¨⊂W`   e. `(⊂2 3)ρ¨⊂W`
c. `2 3ρ⊂¨W`   f. `2 3ρ¨¨W`

2. If `Y←'HIGH' 'AND' 'DRY'` and
   `ALF←'ABCDEFGHIJKLMNOPQRSTUVWXYZ'` what are

a. `ALF⍋⊃Y`   b. `⍋ALF⍋⊃Y`   c. `⍋¨(⊂ALF)⍋¨Y`

3. An experiment consists of rolling a die and counting the number of throws necessary to observe the first 6. Simulate the result of repeating the experiment three times.

4. This exercise is about a titling function PRT3D for rank three arrays and provides in its first line a practical demonstration of partial enclosure. Suppose

```
A←100×⍳2
B←10×⍳3
C←⍳4
ρA21←A∘.+B∘.+C
```
`2 3 4`

Titling consists of two parts per dimension, viz. a descriptor and a vector of individual headings. For example, for the array D the planes could be labelled

   `AAA= 100 ,  AAA= 200`

where the descriptor is `AAA=` and the headings are `100` and `200`. The function PRT3D takes as a left argument a six item vector **V** comprising descriptor/headings for planes, rows and columns respectively.

```
[0]    Z←L PRT3D R;PLA;ROW;COL
[1]    Z←⊂[2 3]A
[2]    Z←' ',[1]¨' ',[2]¨Z
[3]    PLA←L[1],¨2⊃L
[4]    ROW←'\',L[3],4⊃L
[5]    COL←L[5],6⊃L
[6]    Z←(⊂ROW),¨(⊂COL),[1]¨Z
[7]    Z←PLA,[1.5]Z
[8]    Z←,[⍳0]Z
```

```
    TITLES←'AAA=' A 'BBB=' B 'CCC=' C

    TITLES PRT3D A21
AAA= 100

    \ CCC=   1   2   3   4
  BBB=
    10       111 112 113 114
    20       121 122 123 124
    30       131 132 133 134


AAA= 200

    \ CCC=   1   2   3   4
  BBB=
    10       211 212 213 214
    20       221 222 223 224
    30       231 232 233 234
```

Problems:

a. Study the function **PRT3D** and attach a short comment to each line.

b. Fill in the following table for the value of the result variable Z following execution of the lines indicated:

	≡	ρ	ρ¨
[1]			
[2]			
[6]			
[7]			
[8]			

c. In line **PRT3D[3]** why is the first item indexed (L[1]) but the second item picked (2⊃L)?

d. Suppose all the headings are to be character vectors, e.g.

```
PLA1  PLA2
ROW1  ROW2  ROW3
COL1  COL2  COL3  COL4
```

What changes are needed in (i) **TITLES**? (ii) **PRT3D**?

5. The Pascal Triangle of size N consists of the non-zero entries in the outer product (ιN)∘.!ιN. Write functions **PASCAL** and **CENTER** to achieve the following displayed versions:

```
        PASCAL  8
1
1 1
1 2 1
1 3 3 1
1 4 6 4 1
1 5 10 10 5 1
1 6 15 20 15 6 1
1 7 21 35 35 21 7 1
1 8 28 56 70 56 28 8 1

        CENTER PASCAL  8
              1
             1 1
            1 2 1
           1 3 3 1
          1 4 6 4 1
         1 5 10 10 5 1
        1 6 15 20 15 6 1
       1 7 21 35 35 21 7 1
      1 8 28 56 70 56 28 8 1
```

2.1.4 Index with Each

The basic structure for the **index** function is

    I⎕R

where R is a data array and I a set of selection indices. Compared with bracket indexing the **index** function has the advantage of requiring no semicolons. This allows a defined operation to select items via indexing from an array of arbitrary rank. The length of the left argument must match the rank of the right, that is

    (ρ,I) ↔ ρρR

The left argument of **index** may be an array of depth two or less. This permits multiple indices along each of the dimensions of R. Thus for a rank 2 array R

    (1 2)(3 4)⎕R

is equivalent to R[1 2;3 4]. In words this expression selects all the items from R which lie in rows one and two, and also in columns three and four. In conjunction with the **each** operator, the **index** function permits *scatter indexing* which is the process of selecting items at will from an array. With scatter indexing the sets of indices are regarded as separate and independent. For example to select just the item in the first row and second column and the item in the third row and fourth column, the same left argument is used as in the previous example but it is necessary to make the **index** function the operand of **each** with a scalarized right argument thus:

    (1 2)(3 4)⎕¨⊂R

Pictorially this can be shown:

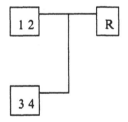

which is equivalent in bracket indexing to (R[1;2])(R[3;4]).

On the other hand to find the item in row 1 and column 2 of *several* matrices the index left argument must be scalarized in conjunction with the application of **each**. Pictorially the situation is

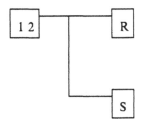

and the appropriate APL2 expression is:

    (⊂1 2)⌷¨R S

which is equivalent to (R[1;2])(S[1;2]). A depth-two index, e.g.

    (⊂(1 2)(3 4))⌷¨R S

is equivalent to:

    (R[1 2;3 4])(S[1 2;3 4])

Now go one stage deeper.

    1 3⌷¨⊂'ABC'

means form a two-letter word from the first and third letters of the alphabet
'ABC'. To form two words, say a one-letter word and a two-letter word, define

    [0]   Z←L SEL R
    [1]   Z←L⌷¨⊂R

and again scalarize the alphabet but this time as right argument to the derived
function SEL¨, e.g.

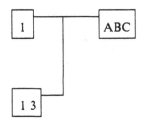

        (1(1 3))SEL¨⊂'ABC'
    A AC

To form two one-letter words requires an explicit use of **ravel** on account of
the fact that enclosure of simple scalars does not increase depth.

        ,¨1 3 SEL'ABC'
    A C

Axis specification may be applied to the **index** function, and the resulting construct

    L⌷[I]R

is called **index with axis**. Here is an example:

DISPLAY M11

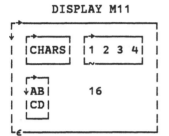

DISPLAY 2 2 1⍴[1]¨⊂M11

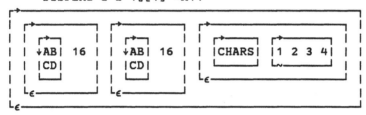

The axis qualifier can be thought of as an operator whose derived function is
⍙[1].

For Z←L⍙R the following identity holds

    ⍴Z ↔ ⊃,/⍴¨L

and for Z←L⍙[I]R, ⍴Z is the shape of R with the Ith item replaced by ⊃,/⍴¨L.

2.2 Extensions to the Slash Operator

A full discussion of how reduction, scan and inner and outer product have been
refined to deal with nested arrays is given in Chapter 5. However two straightforward extensions of reduction merit immediate attention.

2.2.1 Replicate

In APL2 the **slash** operator may take as its left operand either a rank one or
zero data array, or any function producing such a result. With data **V** as the
operand, the derived function **V/** is called **replicate**. This enhances what was
previously called **compression** by allowing the data operand to consist of a
simple scalar or vector of integers. When the operand is a vector of just zeros
and ones it may still be called **compression** as in:

    1 0 1/'ABC'
AC

With non-negative integers the vector operand acts as a mask on the data argument, and the integer determines the number of times the matching data item is replicated. Thus:

```
      3 0 2/'ABC'
AAACC
```

Negative integers in the left argument result in the indicated number of fill items (see Section 1.3.2) being inserted in the designated position, e.g:

```
      1 ¯2 3 0 2/ι4
1 0 0 2 2 2 4 4
```

```
      DISPLAY 0 ¯2 2 0 3/('A'5) 'A' 7 'B'
```

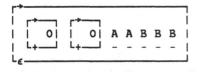

The relationship between the left operand L and the right argument R is either:

1. L is a non-negative scalar in which case it applies to all the items of R, or
2. the number of non-negative items of L must match the shape vector item of R corresponding to the dimension in which replication is to occur. Formally if L/[I]R is valid then the following identity holds:

$$(+/¯1≠×L) ↔ I⊃ρR$$

Illustration : The conjunction IF

Throughout the remainder of this book the function:

```
[0]    Z←L IF R
[1]    Z←R/L
```

will be used so that where there are branches in functions

```
      →L1 IF condition
```

mirrors the English words "branch to L1 if ..." .

Illustration : Multiple copies of matrix rows

Obtain one copy of the second row and two copies of the fifth row of a matrix.

```
      M21←5 3ρ'ANDBOYCANDADEAT'
      0 1 0 0 2/[1]M21
BOY
EAT
EAT
```

It should be emphasized that the slash in **replicate** is a monadic operator symbol and so a vector **v** to its left is an *operand* and the combination **v/** is a derived function. This has a practical consequence which is illustrated by the following sequence:

```
      1 0 1/'ABC'
AC
      1 0 1/'ABC' 'DEF' 'GHI'
 ABC GHI
      1 0 1/¨'ABC' 'DEF' 'GHI'
 AC DF GI
      (1 0 1)(1 1 0)(0 1 1)/¨'ABC' 'DEF' 'GHI'
DOMAIN ERROR
      (1 0 1)(1 1 0)(0 1 1)/¨'ABC' 'DEF' 'GHI'
      ∧                       ∧
```

The DOMAIN ERROR occurs because the operator / is evaluated *before* the operator **each** and the left operand of / must be simple. The intention was presumably to achieve

```
      (1 0 1/'ABC')(1 1 0/'DEF')(0 1 1/'GHI')
```

using **each**. However in order to apply each simple vector to the corresponding component of the right argument, the operand has to be made into an argument by creating a defined function, e.g.:

```
[0]    Z←L COMPRESS R
[1]    Z←L/R

      (1 0 1)(1 1 0)(0 1 1)COMPRESS¨'ABC' 'DEF' 'GHI'
 AC DE HI
```

In the special case however where the left operand L contains just one simple item the derived function L/ is evaluated first and so **each** can be applied to it to give e.g.

```
      1/¨(⍳2)(⍳3)
 1 2  1 2 3

      1 0 2/¨(⍳3)(3+⍳3)
 1 3 3  4 6 6
```

When reduction is applied to non-scalar functions such as ρ and , the intermediate results in general possess structure and so must be enclosed to obtain the correct rank. For example

```
      ρ/2 3
```

has the value 2ρ3, that is 3 3, but rank reduction demands that the final result is ⊂3 3. Similarly

```
      ,/2 3
```

results in ⊂2 3. Thus both ρ/ and ,/ increase depth as the cost of reducing
rank. Similar considerations apply to **scan** and **expand**.

Illustration : Avoiding Blanks in List Lengths

(ι0)s can sometimes be deceptive on account of their "invisibility" in output,
for example

```
      ρ¨(2 2ρι4)(3)(4 5 6 7)
2    4
```

One way of overcoming this by turning scalars into one-item vectors is:

```
      ρ¨1/¨(2 2ρι4)(3)(4 5 6 7)
2 2   1   4
```

2.2.2 Dyadic Reduction

The derived function **reduction** has a dyadic form called **n-wise reduction**
S F/[I] R.   S, which must be a scalar integer, defines a "window" of consecutive items to which reduction applies. The window moves along axis I one
position at a time until all items of R have been covered. For example, a vector
of consecutive pairs of items is given by:

```
      2,/ι4
1 2   2 3   3 4
```

The expression (ι4),/¨⊂ι4 produces a vector of vectors containing the 1, 2, 3
and 4 tuples while 5,/ι4 defines the prototype of ι4.

If the left argument L of **n-wise reduction** applied to a vector is a negative
integer the items within the window are reversed before reduction is applied, so
that (-L),/V is equivalent to ΦL,/ΦV.

Illustration : Reversing scans

+\V can be reversed by ¯2 reduction, that is

```
      V ≡ ¯2-/0,+\V
```

is true for all numeric vectors V. The analogous formula for reversing ×\ is

```
      V ≡ ¯2÷/0,×\V
```

which is true for all numeric vectors V which do not contain a zero. In the
binary domain =\ and ≠\ are reversed using the identities

```
      V ≡ ¯2=/1,=\V      and      V ≡ ¯2≠/0,≠\V
```

A fuller account of scans can be found in Section 5.5.3.

Integers greater than $|1+\rho V$ in the left argument of **n-wise reduction** result in a
DOMAIN ERROR. The left argument may be zero, e.g.

        0,/ι4

is a vector of five ⊂ι0s. This example demonstrates the consistency of APL2 in
dealing with edge conditions by fulfilling the identity

        (S+ρS F/V) ↔ (1+ρV)

Note: 0,/V may give a DOMAIN ERROR on some APL2 implementations.

Illustration : Partitioning a Record into Fields

Suppose a record RC read from a file is 40 characters in length and contains
four fields of lengths 10, 15, 8, and 7. The following expression uses **n-wise
reduction** to split RC into four fields:

        FW←10 15 8 7
        (FW/ιρFW)⊂RC

Exercises 2c

1. Create a function **Compress** which is a modification of the function COM-
PRESS in Section 2.2.1 and which accepts an axis specification as part of an
argument, e.g.:

```
        M M1
ABC     PQR
DEF     STU
GHI     VWX
        (1 0 1)1 Compress M
ABC
GHI
        (⊂(1 0 1)2) Compress¨M M1
AC     PR
DF     SU
GI     VX
```

2. What are the values of

        a.   2-/10 5 2 12 6   b.   ‾2-/10 5 2 12 6
        c.   2ρ/2 3 4   d.   ‾2ρ/2 3 4?

3. a. State in words the result of the expression ((2×ρV)ρ0 ‾1)/V. (Consider
as a test case V←(5 'A') (ι3) 'B' 23).

b. Write an expression which will replace each item in a vector **v** with its own prototype. (Hint - Use the COMPRESS function in Section 2.2.1).

4. How many scans other than +\, ×\, =\ and ≠\ can you reverse using ⁻2 **n-wise reduction** in the style of Section 2.2.2?

5. If **v** is a simple vector and **F** is a primitive scalar dyadic function what does the following expression define :

```
F/¨(⍳⍴V)↑¨⊂V    ?
```

6. a. Write a recursive function DTB to delete trailing blanks (but no others) from a character vector.

   b. How else could this be could this be achieved in APL2?

   c. How would you use DTB to remove trailing blanks from every word in a vector of words **VW**?

7. a. For a simple numeric vector **v** write an expression for the product of all consecutive pair-wise sums of items, e.g. if **v** is ⍳5 the result is 3×5×7×9 = 945.

   b. Describe the string

```
∈3 1 4ρ¨'ABC'
```

in terms of just one primitive function or operator.

8. Write a function FIND which behaves like ∊ (see Section 1.6.3) except that it returns a 1 in the position corresponding to *every* matched character, e.g.

```
     'CAT' FIND 'BATTY CATS SCATTER DUCATS'
0 0 0 0 0 0 1 1 1 0 0 0 1 1 1 0 0 0 0 0 0 1 1 1 0
```

9. In the third illustration of Section 1.6.3 how would you amend the expression to find all occurrences of the pattern

```
      PAT
1 X
X 1
```

where **X** stands for "don't care," that is the bit concerned may be either a **0** or a **1**?

Summary of Functions used in Chapter 2

Section 2.1.1
MIDPT   mid-points in Euclidean geometry

Exercises 2b
PRT3D   titles three dimensional matrix
PASCAL   Pascal's triangle
CENTER   centers rows of character matrix

Section 2.1.4
SEL   multiple scatter indexing

Section 2.2.1
IF   conditional branching
COMPRESS   functional form of compress operator

Exercises 2c
DTB   deletes trailing blanks
FIND   enhancement of primitive function **find**

3
Elementary Data Structuring

The objective of this chapter is to demonstrate the effectiveness of APL2 in dealing with relatively straightforward commercial and financial programming situations based on real applications. There are two main sections each of which is given in the form of an extended illustration with narrative. The final exercises try to show how rapidly a reasonably substantial application can be built up from scratch.

3.1 Example 1. Product Stocks

This example is about stocks of five components X801 to X805 which are bought from three countries, JAPAN, TAIWAN and HONGKONG. The relevant stocks for each component are given below (JAPAN does not deal with X804 or X805 and HONGKONG does not deal with X805 or X802):

```
JAPAN+45 75 15
TAIWAN+35 75 15 45 95
HONGKONG+35 0 55 15
```

With such a data organization a variety of questions can be asked, e.g. what is the total number of JAPAN components?

```
+/JAPAN
135
```

Each allows all countries to be processed simultaneously:

```
+/¨JAPAN TAIWAN HONGKONG
135 265 105
```

The total stock of components is

```
+/∈JAPAN TAIWAN HONGKONG
505
```

or if

```
STOCKS+JAPAN TAIWAN HONGKONG
```

then

```
     STOCKS
 45 75 15   35 75 15 45 95   35 0 55 15
```

and

```
     +/¨STOCKS
135 265 105
```

and

```
     +/∊STOCKS
505
```

To find the total stocks by components two options are available. Either country vector is padded out with zeros so that the component items are in matching positions:

```
     +/5↑¨STOCKS
115 150 85 60 95
```

or ⊃s are used to create a matrix within which the padding takes place automatically and then sum down the columns:

```
     +/⊃STOCKS
115 150 85 60 95
```

Now enter the costs of the various components ...

```
     COSTS←(39 19 29)(35 15 29 15 45)(25 15 19 12.5)
```

Suppose X802 for JAPAN was in error. The X802 cost component for JAPAN is set to the correct value of 15 by:

```
     (1 2⊃COSTS)←15
     COSTS
 39 15 29   35 15 29 15 45   25 15 19 12.5
```

The total cost of inventory by country is:

```
     COSTS+.×¨STOCKS
3315 7735 2107.5
```

and the total cost of inventory is:

```
     +/∊COSTS×STOCKS
13157.5
```

By establishing the names of the countries as a variable ..:

```
     CNTRIES←'JAPAN' 'TAIWAN' 'HONGKONG'
```

.. the cost of inventory for each country can be displayed as

```
     CNTRIES,¨COSTS+.×¨STOCKS
 JAPAN 3315   TAIWAN 7735   HONGKONG 2107.5
```

Suppose all prices are to be marked up by 80% :

```
      COSTS×1.8
70.2 27 52.2   63 27 52.2 27 81   45 27 34.2 22.5
```

Each entry may have different markups:

```
      PRICES←COSTS×1.6 1.7 1.8
      PRICES
62.4 24 46.4   59.5 25.5 49.3 25.5 76.5   45 27 34.2 22.5
```

The resulting net markups for each country/component combination are:

```
      NETMU←STOCKS×PRICES-COSTS
      NETMU
1053 675 261   857.5 787.5 304.5 472.5 2992.5   700 0 836 150
```

.. and the net markup for the entire stock is:

```
      +/∈NETMU
9089.5
```

The average percent markup is:

```
      ⌊.5+100×(+/∈NETMU)÷+/∈COSTS×STOCKS
69
```

.. and the biggest net markup in each country is given by:

```
      ⌈/¨NETMU
1053 2992.5 836
```

By establishing the component names as CNOS

```
      CNOS←'X801' 'X802' 'X803' 'X804' 'X805'
```

the maxima may be stated by component name:

```
      CNOS[NETMU⍳¨⌈/¨NETMU]
X801 X805 X803
```

Instead of viewing STOCKS as a vector of vectors:

```
      STOCKS
45 75 15   35 75 15 45 95   35 0 55 15
```

it may be viewed in tabular form:

```
      ⊃STOCKS
45 75 15  0  0
35 75 15 45 95
35  0 55 15  0
```

or...

```
      ⊃[1]STOCKS
45 35 35
75 75  0
15 15 55
 0 45 15
 0 95  0
```

Note that missing fields are automatically filled with zeros.

The tabular representation of **STOCKS** may be prefaced by country labels:

```
        CNTRIES,⊃STOCKS
JAPAN     45 75 15  0  0
TAIWAN    35 75 15 45 95
HONGKONG 35  0 55 15  0
```

... and column headings may be added:

```
      (' ',CNOS),[1]' ',[1]CNTRIES,⊃STOCKS

            X801 X802 X803 X804 X805

JAPAN        45   75   15    0    0
TAIWAN       35   75   15   45   95
HONGKONG     35    0   55   15    0
```

So far all the titles have been added in interactive mode. It could be embedded in a function such as

```
        ∇Z←L TOPS R
[1]     Z←(' ',L),[1]' ',[1]R∇

      CNOS TOPS CNTRIES,⊃STOCKS

            X801 X802 X803 X804 X805

JAPAN        45   75   15    0    0
TAIWAN       35   75   15   45   95
HONGKONG     35    0   55   15    0
```

Suppose the component names are too long to make the table look nice. One possibility would be to list the names vertically which leads to the following alternative presentation :

```
      (⊃[1]CNOS)TOPS CNTRIES,⊃STOCKS

            X  X  X  X  X
            8  8  8  8  8
            0  0  0  0  0
            1  2  3  4  5

JAPAN     45 75 15  0  0
TAIWAN    35 75 15 45 95
HONGKONG 35  0 55 15  0
```

Or perhaps there should be more spaces in the original report ...

```
      (6 0⍕CNOS)TOPS CNTRIES,⊃(⊂6 0)⍕¨STOCKS

            X801   X802   X803   X804   X805

JAPAN         45     75     15
TAIWAN        35     75     15     45     95
HONGKONG      35      0     55     15
```

Because ⍕ is used this version shows blanks for items which are not present.

Summary reports are obtained by:

```
      'TOTAL INVENTORY',+/∊STOCKS
TOTAL INVENTORY 505

      (⊃CNTRIES),+/¨STOCKS
JAPAN      135
TAIWAN     265
HONGKONG 105

      (⊃CNOS),+/⊃[1]STOCKS
X801 115
X802 150
X803  85
X804  60
X805  95
```

To view all this information collectively use:

```
      A←(⊃CNOS),+/⊃[1]STOCKS
      B←(⊃CNTRIES),+/⊃STOCKS
      C←'TOTAL INVENTORY',+/∊STOCKS
      A B C
 X801 115    JAPAN      135    TOTAL INVENTORY 505
 X802 150    TAIWAN     265
 X803  85    HONGKONG 105
 X804  60
 X805  95
```

... or vertically:

```
      ,[⍳0]A B C
X801 115
X802 150
X803  85
X804  60
X805  95

JAPAN      135
TAIWAN     265
HONGKONG 105

TOTAL INVENTORY 505
```

Finally for a full spreadsheet type of report:

```
      HEAD←CNOS,⊂'TOTS'
      BODY←CNTRIES,(⊃STOCKS),+/⊃STOCKS
      FEET←(⊂'TOTALS'),(+/⊃[1]STOCKS),+/∊STOCKS
      HEAD TOPS BODY,[1]FEET
```

```
              X801 X802 X803 X804 X805 TOTS

JAPAN           45   75   15    0    0  135
TAIWAN          35   75   15   45   95  265
HONGKONG        35    0   55   15    0  105
TOTALS         115  150   85   60   95  505
```

```
      ρHEAD TOPS BODY,[1]FEET
6 7
```

Typically this would be embodied in a function that allowed the report to be run repeatedly with only a change in STOCKS.

```
[0]    Z←HR REPORT R;HEAD CNTRIES FEET
[1]    ⍝HR: Two item vector of row and column titles
[2]    (HEAD CNTRIES)←HR
[3]    HEAD←(''),HEAD,⊂'TOTS'
[4]    BODY←CNTRIES,(⊃R),+/⊃R
[5]    FEET←(⊂'TOTALS'),(+/⊃[1]R),+/∊R
[6]    Z←HEAD TOPS BODY,[1]FEET
```

The above report is then produced by

```
      (CNOS CNTRIES)REPORT STOCKS
```

Of course a good spreadsheet reporting system could have done as much up to this point, however with APL2 this is only a beginning. APL2 doesn't just format reports - it is a powerful general-purpose language which as well as doing the simple sums above could equally well have performed complex statistical functions tailored to the user's special needs.

Exercises 3a

1. For the example above find the average selling prices for each of the five components for two different definitions of "average" namely

  a. a simple average of the selling prices in countries which hold stocks of the component;
  b. an average weighted by quantities in stock.

2. Order the components by decreasing profitability within each country. First obtain the answer as indices, then translate these into component names.

3. A Cash Register system

Many cash registers not only print out the cost of each item but also the name of the item. In order to do this assume that the system has a **STOCK** database. Suppose each entry in this database consists of (inventory number) (product name) (unit amount)(cost per unit). For example:

```
SCREWS←211 'THREADIES' 1000 1.98
SPANNERS←312 'FLATONES' 1 1.09
PLUGS←654 'LOTSAVOLTS' 2 1.55
STOCK←SCREWS SPANNERS PLUGS
```

Define a function **RECEIPT** which will accept a vector of inventory numbers and the name of the stock database, and produce a matrix listing of the product names, unit amounts and costs per unit that match the inventory numbers.

3.2 Example 2. Optimizing Rental Charges

This is an example of a financial application using the ideas of probability and discounted cash flow. A business offers both lease and outright purchase options on its products, and requires to determine "multipliers" where a multiplier is defined as

$$\frac{\text{outright purchase price}}{\text{monthly rental charge}}$$

The customer's decision to buy or rent is determined by two factors, namely his perception of the lifetime of the product and his view of the discounted present value of future rental payments. These can be combined in a discounted cash flow calculation

```
[0]    Z←L PCENTFOR R
[1]    Z←+/÷(1+.01×L÷12)∘.*ι12×R
```

which results in the total number of monthly rentals discounted at L% which will be paid over a period of R years. This is equivalent to the maximum number of monthly rentals which the rational customer would be prepared to pay for outright purchase. For example with cash discounted at 7%, the equivalent number of monthly rentals paid over four years at present values is given by

```
      2▪7 PCENTFOR 4
41.76
```

For a given product at a given moment in time the perception of discount rate and lifetime varies over a range of existing and prospective customers, and for modelling purposes it is assumed that both can be described in terms of estimated frequency distributions. The structure of customer lifetime perceptions will also vary for different products according to their rate of depreciation, e.g. television sets wear out gradually, whereas software becomes obsolete in a more sudden fashion.

On the financial side, discounted rates for the value of future money varies for different customers even at the same moment in time.

Suppose that the vectors **DRATE** and **LIFEX** describe possible discount rates and lifetime expectancies, and that the vectors **DLIST** and **LDIST** represent the proportions of customers holding these beliefs.

```
      DRATE←7 9 11 13
      LIFEX←4 5 6

      DDIST←.2 .4 .3 .1
      LDIST←.25 .5 .25
```

i.e. 20% of customers discount future money at 7%, 40% at 9% and so on, while 25% estimate product lifetime as four years, etc.

The joint distribution of the discount rate DRATE and the life expectancies
LIFEX is the outer product

```
        2⍕DRATE∘.PCENTFOR LIFEX
41.76 50.50 58.65
40.18 48.17 55.48
38.69 45.99 52.54
37.28 43.95 49.82
```

and the matching joint distribution of expectancies is the outer product

```
        3⍕DDIST∘.×LDIST
.050 .100 .050
.100 .200 .100
.075 .150 .075
.025 .050 .025
```

Suppose now that the supplier fixes the multiplier at 50 say. If all customers
are assumed to behave rationally what proportion will opt for purchase? Call
the two tables above MTAB and PTAB standing for multipliers table and proportions table respectively and define a function CUSPRO:

```
        MTAB←DRATE∘.PCENTFOR LIFEX
        PTAB←DDIST∘.×LDIST
```

```
[0]    Z←L CUSPRO R;MTAB;PTAB
[1]    ⍝L: a 2 item vector - multipliers table and proportions table
[2]    ⍝R: a multiplier
[3]    ⍝Z: %opt to purchase
[4]    (MTAB PTAB)←L
[5]    Z←+/(R<,MTAB)/,PTAB
```

The above question for a multiplier of 50 is then answered by

```
        (MTAB PTAB) CUSPRO 50
0.325
```

Now extend this to a *range* of multipliers.

```
        (⊂MTAB PTAB) CUSPRO¨ 40 45 50
0.9 0.7 0.325
```

What if the distribution of customer lifetime perceptions varies according to
whether the customer is a new or an existing renter? To accommodate this
adjust LIFEX to include smaller lifetimes and LDIST to be a two-item vector
where the first item reflects the distribution of a new customer and the second
item the distribution of an existing renter.

```
        Lifex←⍳6
        Ldist←(0 0 0 .25 .5 .25)(.1 .2 .3 .2 .2 0)
```

The two outer products require adjustment, in particular the one which multi-
plies the distributions since what is needed is two separate tables corresponding
to each of the two discrete distribution items of Ldist:

```
      Mtab+DRATE•.PCENTFOR Lifex
      Ptab+(⊂DDIST)•.×¨Ldist
```

Since `Ptab` is now a vector comprising two items, each of them a table, it is the *derived* function COMPRESS¨ (see Section 2.2.1) which has to be applied to each of them, and similarly for , ¨ and +/¨. This requires that CUSPRO be rewritten as

```
[0]   Z+L Cuspro R;Mtab;Ptab
[1]   (Mtab Ptab)+L
[2]   Z++/¨(⊂R<,Mtab)COMPRESS¨,¨Ptab
```

so that its result is now a two-item vector with each item corresponding to one of the customer classes, new or existing:

```
      (Mtab Ptab)Cuspro 50
0.325 0.04
```

To investigate a range of multipliers apply **each** as before:

```
      (⊂Mtab Ptab)Cuspro¨40 45 50
 0.9 0.32   0.7 0.18   0.325 0.04
```

or if the result is required as a table:

```
      ⊃(⊂Mtab Ptab)Cuspro¨40 45 50
0.9   0.32
0.7   0.18
0.325 0.04
```

To recap so far, the above table contains the proportions of customers opting for purchase. The (implicit) row headers are the multipliers set by the supplier, and the columns relate to new and existing customers respectively.

Given this information together with a forecast of the numbers of new and existing customers, the supplier may now calculate his expected revenues for different multipliers. Suppose that he has done this and has produced a five-item forecast revenue vector REV:

```
      REV+3300  4000  5300  6400  7000
```

(It is true that the forecast of numbers of new and existing customers is likely to depend on the multiplier but this complexity is ignored for the time being.)

The supplier as well as his customers has a perception concerning the discounted value of money. How does this allow the supplier to maximize his revenue? The following function returns net present value:

```
[0]   Z+L NPV R
[1]   ⍝L: discount rate as a percentage
[2]   ⍝R: vector of amounts
[3]   Z+R÷(1+.01×L)*¯1↓⍳⍴R+,R
```

The value of a single sum discounted for different rates for one year is given by

```
      (⊂10 12)NPV 100
90.909 89.286
```

and the outer product

```
      10 12∘.NPV 100 200 300
  90.909  181.82  272.73
```

gives the value of several sums discounted for one year at different rates.

For a given revenue estimate vector **REV** the supplier can now discount the first item over one year, the second over two years and so on to give:

```
      10 NPV REV
3000 3305.8 3982 4371.3 4346.4
```

as the projected revenue discounted at 10%.

Applying **each** to the revenue vector allows this calculation to be performed for several discount rates, e.g.

```
      +/¨10 12 NPV¨⊂REV
19005 17947
```

returns the total revenues with discounting at 10% and 12% respectively. Now at last the supplier is in a position to examine **simultaneously** his returns under a variety of assumptions and to make what he believes to be an optimum decision in setting the multiplier.

Exercises 3b

1. Given **MTAB** and **PTAB** as defined above, estimate the value of the multiplier at which 50% of customers opt for purchase? (Hint: order the items in **,PTAB** according to the corresponding values in **,MTAB**.)

2. Obtain a table of discount rates 10% and 12% versus **several** revenue projections, e.g.

```
REV←3300 4000 5300 6400 7000 (as in the text above)
REV1←5ρ3300
REV2←3300×(1.05)*0,ι4
```

3. The staff of a department starts a savings bank, and records the individual transactions of members in a numeric nested vector **BANK** which is structured hierarchically as follows:

```
BANK
  - MEMBER RECORD
     - MONTH
        - TRANSACTIONS
```

BANK is a vector of **MEMBER RECORDS**. A **MEMBER RECORD** is in turn a vector of **MONTHS** of positive or negative **TRANSACTIONS** where a deposit is indicated by a positive number and a withdrawal by a negative number. At the transaction level a deposit is indicated by a positive number, and a withdrawal by a negative number. A **MONTH** is a vector of transactions, a **MEMBER RECORD** is a vector of months, and **BANK** is a vector of member records.

A typical instance of **BANK** after one such bank had been operating for two months with four members is

```
BANK←((20 ¯5 10 ¯5)(4 ¯2 ¯7))((¯10 25)(16 3))
                         ((5 ¯9 ¯2)(6 ¯3 ¯3))(25(ι0))
```

Give APL2 expressions which answer the following for this or any similarly structured bank:

a. How many members has the bank?

b. For each member over the entire period, what are
   (i) his net deposits?
   (ii) his total deposits?
   (iii) his total withdrawals?

c. What is each member's sequence of monthly balances over the period?

d. What are the bank's net deposits by month over the period?

e. What is the net amount on deposit with the bank at the end of the period?

4. Last Trades.
Consider the following portion of data from the stock exchange:

```
MMM   3:25   95
T     3:27   36.5
GM    3:31   43
MMM   3:33   42.75
IBM   3:45  102.25
IBM   3:57  102.125
GM    4:02   43.125
GM    4:04   43.375
IBM   4:04  102.25
T     4:05   36.75
IBM   4:12  102.5
```

Assume it is structured as a nested three column matrix **STP** in which each row represents a trade on the stock exchange, the first column is the trading symbol, the second the time of the trade and the third the trading price.

Problems:

a. Create a function **LAST_TRADE** to find the last trade of each stock.

b. Modify the function to another function **STK_LAST_TRADE** which finds the last trade of a given stock.

c. Enhance the program to return a message if the stock is not traded.

d. Modify **LAST_TRADE** to a function **TIM_LAST_TRADE** which returns the last trade of each stock after a given time.

Summary of Functions used in Chapter 3

Section 3.1
REPORT   formatted report

Exercises 3a
RECEIPT   builds cash register receipt

Section 3.2
PCENTFOR   discounted cash flow calculation
CUSPRO   customer proportion opting to purchase
NPV   net present value

Exercises 3b
LAST_TRADE   last trade of each stock
STK_LAST_TRADE   last trade of given stock
TIM_LAST_TRADE   last trade after given time

4
Using Functions and Arrays

Chapters 1 and 2 discussed functions and operators in relative isolation. Chapters 4 and 5 are about the *interactions* of various functions and operators and thus provide a study in greater depth of the features of APL2 which are most intimately connected with nested arrays and the associated operations of **enclose**, **disclose** and **each**. Although the basic concepts are few a new perspective is required in order to acquire fluency in application. There is an analogy to the mental leap needed to move from thinking in two dimensional geometry to thinking in three dimensions. Data objects in first-generation APL possess data and structure where structure is synonymous with shape. In APL2 structure is given the additional aspect of *depth*, thereby releasing the user from the shackles of rectangular data structure and thus making it possible to model data structures of almost indefinite complexity. The price of the flexibility afforded by the combination of data, depth and shape is that the simultaneous control of all three is a skill which has to be consciously acquired through practice in order to exploit the great programming versatility which nested arrays afford.

4.1 Cross-sections, Picking and Indexing

The distinction between *items* on the one hand and *cells* on the other, that is objects *containing* items, is crucial to understanding nested arrays. In general, indexing creates *cross-sections* of arrays, and for a two-item vector V this can be pictured

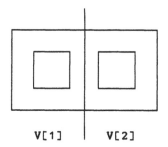

whereas **pick** *penetrates* arrays, thereby reducing depth.

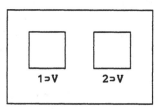

Thus if **V** is a vector, then contrary to what one might intuitively suppose, **V[1]**
is *not* the first item of **V**, but rather it *contains* the first item of **V**. Enclosure
thus provides a *container* for data and indexing selects from a rectangular array
of containers.

   Disclose on the other hand removes a container, allowing functions to be performed on its contents. The following vector identity reinforces this point:

$$V[I] \quad \equiv \quad \subset I \supset V$$

(cf. post- and pre-brackets in Section 1.3.3). For example suppose

        V←'GO' 'TO' 'BED'

2⊃V is the two-item vector 'TO' whereas P[V], or equivalently 2⌷V, is ⊂2⊃V,
that is the **enclose** of 'TO'. Thus

        V[2]≡⊂'TO'
1
        (2⊃V)≡⊂'TO'
0

Also

        V[3]=⊂'BED'
1 1 1
        (3⊃V)=⊂'BED'
 1 0 0   0 1 0   0 0 1

The last expression answers the nine separate questions

        'B'='B'   'B'='E'   'B'='D'
        'E'='B'   'E'='E'   'E'='D'
        'D'='B'   'D'='E'   'D'='D'

A visually similar relationship exists between the argument and result of an **each**-derived function. This may be expressed:

If F is a monadic function and Z←F¨R where R is a vector then for all valid I

$$Z[I] \quad \leftrightarrow \quad \subset F \supset R[I]$$

The sequence "enclose-function-disclose" is of such frequent occurrence that the above statement will be called the "**each** rule." Since **disclose** is the inverse of **enclose** the above equivalence can be seen as a manifestation of the formula GFG^{-1} which pervades mathematics, and linear algebra in particular. Z and R have the same shape, that is

$$\rho Z \quad \leftrightarrow \quad \rho R$$

and the following identities also hold

```
↑F I⌷R        ↔      F↑I⌷R
F¨R1 R2       ↔      (F R1)(F R2)
```

For dyadic F with

```
Z←L F¨R
```

the corresponding form of the **each** rule is

$$Z[I] \leftrightarrow \subset(\supset L[I])F \supset R[I]$$

or equivalently

$$I⌷Z \leftrightarrow \subset(\supset I⌷L)F \supset I⌷R$$

Here the **disclose** step is applied simultaneously to both arguments. The analogous identities are

```
↑I⌷,Z       ↔      (↑I⌷,L)F↑I⌷,R
L1 L2 F¨R1 R2 ↔     (L1 F R1)(L2 F R2)
```

Contrast this with the function **pick** for which the corresponding rules are

$$I \supset Z \leftrightarrow F \ I \supset R$$

$$I \supset Z \leftrightarrow (I \supset L)F \ I \supset R$$

The vector rules for L F¨ R are readily extendible to arrays of higher rank.
Here are the considerations involved in applying the **each** rule to evaluate e.g.

```
V41←((2 2)(3 1))ρ¨((ι4)'ABC')
```

Each argument of ρ¨ is a two-item vector, and so the **each** rule says that in order to obtain the leading item of the result vector 1⊃ must be applied to both arguments to obtain 2 2 on the left and (ι4) on the right. Now ρ is applied and the result enclosed and placed in its proper cell in the result. For the second item do the same with 2⊃ to give a result

DISPLAY V41

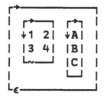

4.1.1 Each and Scalar Functions

Another way to view **each** is in terms of scalar functions where a *scalar* function is defined as one which is applied independently to each item in the case of a monadic function, or between corresponding items of the left and right arguments in the dyadic case. Thus for example + is a scalar function and

    1 2 3 + 4 5 6

can be viewed as shorthand for

    (1+4) (2+5) (3+6)

Regardless of whether or not F is scalar F¨ behaves like a scalar function *one level down* in the data structure of its arguments.

Scalar function behavior also means that scalar extension applies. The two basic forms of scalar extension are illustrated by

	Example	Scalar extension
S F A	2 + 3 4 5	2 2 2+3 4 5
A F S	1 2 3 + 10	1 2 3+10 10 10

The following frequently occurring patterns arise on account of the fact that F¨ is a scalar function:

    S F¨ A B   ↔ (S F A)(S F B)
    A B F¨ S   ↔ (A F S)(B F S)
    A B F¨ C D ↔ (A F C)(B F D)

Another way to evaluate the vector V in the previous example is to observe that

    ((2 2)(3 1)) ρ¨ ((ι4) 'ABC')

is equivalent to

    (2 2ρι4) (3 1ρ'ABC')

When F is itself a scalar function **each** has no role to play since F already penetrates all levels of structure down to the simple items. Thus 1 2+3 4 is identical to 1 2+¨3 4, whereas the following are not identical:

```
      DISPLAY 1 2,3 4
```

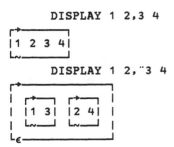

```
      DISPLAY 1 2,¨3 4
```

Scalar extension often comes about through the application of **enclose**, for example:

```
      (⊂3 1) ρ¨ ((ι4) 'ABC')
```

is equivalent to

```
      (3 1ρι4) (3 1ρ'ABC')
```

Because the derived function F¨ behaves like a scalar function one level down, three patterns involving the explicit use of **enclose** tend to arise in expressions, namely:

```
      (⊂A) F¨  B
       A  F¨ (⊂B)
           F¨ ⊂[I] A
```

The first two reflect the scalar-function-array and the array-function-scalar patterns of scalar extension. The third one typically creates a vector for which a function such as **grade-up** or +/ is applied to each item.

The **depth** function further emphasizes the difference between **pick** and indexing in that ≡V[1] is two, and ≡1⊃V is one. Yet another way to look at this is to say that V can be viewed in either of two ways:

(a) as the join of two depth two objects, or
(b) as the enclosure of two depth one objects.

A consequence of this is that care must be taken to distinguish "the first item of an array" from "the first cell of an array." The former implies depth-reduction, the latter not.

Vector notation provides a mechanism for enclosure without explicit use of the **enclose** function as in

```
      'ABC' 'XYZ'▲¨'CAT' 'AXE'
     2 1 3  2 1 3
```

Here L[1] is the scalar ⊂'ABC', R[1] is the scalar ⊂'CAT', so the dyadic form of the **each** rule predicts that Z[1] is ⊂'ABC'▲'CAT'.

4.2 Some Illustrations using Nested Arrays

Sometimes algorithms carry over without change from the non-nested case, e.g.
`((DOCιD)=ιρDOC)/DOC` removes duplicate words from a "document" DOC
which is a vector of character vector "words."

Illustration : Word Search

Test for occurrences of a word in a document.

```
      DOC←'THE ONLY THING TO FEAR IS FEAR ITSELF.'
      DC←(' '≠DOC)⊂DOC        ⍝ partition string into words
      (⊂'FEAR')∈DC            ⍝ is the word present?
1
      (⊂'FEAR')≡¨DC           ⍝ a mask for its occurrences
0 0 0 0 0 1 0 1 0
```

Illustration : Spell Check

Find the words in TEXT which are not in DICT. Thus for spell checking, TEXT
is the text as a vector of words, and DICT is the dictionary, also as a vector of
words.

```
      DICT←'RECEIPT' 'THE' 'THEIR' 'THERE' 'WAS'
      TEXT←'THIER RECIEPT WAS THERE'
      TEXT←(' '≠TEXT)⊂TEXT
      TEXT~DICT
 THIER  RECIEPT
```

Illustration : Enlarging a List of Words

Let LIST be an existing vector of words, for example

```
      LIST←(' '≠LIST)⊂LIST←'BOOK READ THE TO TOO'
```

and

```
      TEXT←'TO' 'READ' 'THE' 'TWO' 'RED' 'BOOKS' 'TOO'
```

Following

```
      LIST←LIST,TEXT~LIST
```

LIST is updated with the words of TEXT not previously in LIST.

Illustration : Vector Merge

Merge two vectors in the sense of taking one item alternately from each. For example:

```
      V42←'PETER '    'PAUL '    'MARY '
      V43←'AND '    'AND '    'BROWN'
      ∊V42,¨V43
PETER AND PAUL AND MARY BROWN
```

Illustration : Random Sentence Building

Given a vector of subjects, a vector of verbs and a vector of nouns, the following function will generate random sentences.

```
[0]    Z←SENTENCE SVN
[1]    ⍝SN:  a three item vector of vectors
[2]    ⍝Z:   a random sentence consisting of subject, verb and noun
[3]    Z←(?⍴¨SVN)⊃¨SVN
```

For example, with

```
      SUBJECTS←'RAY' 'NORMAN' 'JO' 'JEAN' 'DAVID'
      VERBS←'EATS' 'LIKES' 'DISLIKES' 'ENJOYS'
      NOUNS←'FISH' 'OATMEAL' 'APPLES' 'OLIVES' 'SPINACH'

      SENTENCE SUBJECTS VERBS NOUNS
```

might produce

```
JO LIKES OLIVES
```

A common problem when using **each** in APL2 programming is ensuring that corresponding **encloses** and **discloses** are matched correctly. Programming with nested arrays rapidly leads to the discovery that a tiny difference in code can make a large difference in result, and consequently it is important to recognize differences between similar but subtly different expressions. The following exercises emphasize this point, and the solutions illustrate the extra complexity which arises when enclosure and **each** are used together.

Exercises 4a

1. Suppose **V** is the vector 4 5 6. Consider the following set of eight somewhat similar expressions all of which are variations on the theme

```
     2 3⍴V
```

Some are meaningful, some are not. Evaluate those which are and predict the nature of the error in the case of those which are not.

a. `2 3ρ⊂V`   e. `(⊂2 3)ρV`
b. `2 3ρ¨⊂V`   f. `(⊂2 3)ρ⊂V`
c. `2 3ρ¨V`   g. `(⊂2 3)ρ¨V`
d. `2 3ρ⊂¨V`   h. `(⊂2 3)ρ¨⊂V ?`

A detailed discussion of this exercise is given in Appendix A.

2. Here are some similar variations on the theme `'AB','CDE'`. What is the result of executing the following:

a. `'AB',⊂'CDE'`   e. `(⊂'AB'),⊂'CDE'`
b. `'AB',¨⊂'CDE'`   f. `(⊂'AB'),¨⊂'CDE'`
c. `'AB',¨'CD'`   g. `'AB',¨¨'CDE'`
d. `'AB',⊂¨'CDE'`   h. `'AB',¨¨⊂'CDE' ?`

3. a. How would you sort a vector of codes each of which is a mixture of numerics and alphabetics, e.g. `A9, B12, B9, b9, B10, ...` ? Distinguish two cases:

(i) all upper case letters come before any lower case letter;
(ii) all a's in any case come before any b's and so on.

4. Use `⎕AF` to construct the "alphabet" `'AaBbCc ... '`. (see Section 1.5.2 for a description of `⎕AF`.)

5. If `V1` is a vector of words, write an expression which returns a 1 to indicate the occurrence of *any* of the words in `V1` as consecutive characters in a character vector `V2`.

6. Write an expression which returns the index of every occurrence of `'AB*C'` in a character string `V` where
    a. * represents any single character;
    b. * represents any character string of arbitrary length including zero, which does not contain a further `'C'`.

7. Word Analysis
    Suppose you have a variable representing textual data as a simple character vector, (e.g. `GETTYSBURG` representing the Gettysburg address).

    a. How many words does it contain?

    b. How many *distinct* words does it contain (remember to remove punctuation and to change upper case letters to lower case at the start of sentences)?

   c. How many occurrences are there of each of these distinct words, that is
   obtain a concordance of the data with the words sorted in order of frequency of occurrence.

8. What does the following expression do

   ε' ',¨V

given that V is a vector of words?

4.2.1 Further Illustrations using Nested Arrays

Illustration : Catenation of Matrices

A frequent programming task to which APL2 brings a new way of thinking is
that of adjoining two matrices of unequal dimensions, e.g.

```
          DISPLAY"LEFT RIGHT
┌─→─┐    ┌─→────┐
↓PIP│    ↓DICK  │
│TOM│    │ALBERT│
│JO │    └──────┘
└───┘
```

Suppose **LEFT** and **RIGHT** are to be catenated along the first dimension. An
APL2 style algorithm converts each matrix to a vector of rows, catenates them,
and reconstitutes the result as a matrix. A function to do this is

```
[0]    Z←L VCAT R        ⍝ vertical catenation of matrices
[1]    Z←⊃⊃,/⊂[2]"L R
```

```
       LEFT VCAT RIGHT
PIP
TOM
JO
DICK
ALBERT
```

The two **discloses** in **VCAT** relate to two enclosures, one explicit, and the other
implicit arising from `,/` which will be discussed in detail in Chapter 5.

   VCAT as given above requires that both arguments are matrices. If the function is to work with vectors or scalars, `⊂[2]` must be generalized to

```
[0]    Z←PENCL R         ⍝ descalarize and partial enclose
[1]    Z←⊂[⍴⍴R]R←1/R
```

(`1/R` makes R into a one-item vector if it is a scalar, otherwise does nothing - see
Section 1.5.) Therefore rewrite **VCAT** as

```
[0]    Z←L Vcat R
[1]    Z←⊃⊃,/PENCL"L R
```

```
       LEFT Vcat'3'
PIP
TOM
JO
3
```

```
      3 Vcat 4
3
4
      'PIG'Vcat'SHEEP'
PIG
SHEEP
```

Illustration : Partial Enclosure

If it is logical to think of a matrix as a vector of row vectors or as a vector of column vectors, then consider using ⊂[2]M and ⊂[1]M respectively. For example to transform a matrix M with a single 1 in each row, e.g.

```
      0 0 1 0
      1 0 0 0
      0 0 0 1
```

into the column indices the ones by rows use

```
      (⊂[2]M)ι¨1
3 1 4
```

which is arguably more expressive of intention than the regularly used idioms from first-generation APL

```
      M+.×ι¯1↑ρM      and    1++/∧\M≠1
```

or even the APL2 idiom

```
      Mf.ι1
```

where f is any dyadic function. This is discussed in detail in Section 5.5.6.

Illustration : Find the co-ordinates of the 1s in a binary matrix

Given

```
      M41
  1 1 0 0
  0 0 0 1
  1 0 0 0
```

the co-ordinate pairs of the 1s in M41 are

```
      (1,1)   (1,2)   (2,4)   (3,1)
```

One way to obtain these is to use ⊤ :

```
[1]    Z←ONES R
[2]    Z←⊂[1]1+(ρR)⊤¯1+(,R)/ι×/ρR
```

of which a more elegant but less efficient form is:

```
[1]    Z+Ones R
[2]    Z+1+(⊂ρR)T¨¯1+(,R)/ιρ,R
```

⊃ONES M41 displays the co-ordinate pairs, one per row:

```
    ONES M41
1 1   1 2   2 4   3 1
```

If it is desired to retain the row structure in the solution each row of M should be thought of separately as a compression vector on ι4, so following the reasoning of Section 2.1 ι4 should be scalarized as indicated by the following diagram:

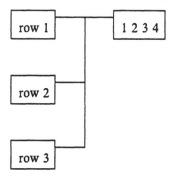

and the function COMPRESS

```
[0]    Z+L COMPRESS R
[1]    Z+L/R
```

used to give

```
      DISPLAY CI+(⊂[2]M41)COMPRESS¨⊂ι4
```

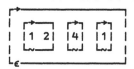

Next each row number is joined with its own set of indices using the function ,¨:

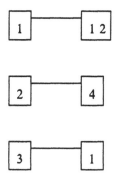

which is achieved by the expression

        DISPLAY(ι3),"¨CI

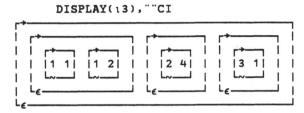

Adding ⊃¨:

        ⊃¨(ι3),"¨CI
  1 1   2 4   3 1
  1 2

gives what is perhaps a more satisfactory display, or more generally

        ⊃¨(ι↑ρM),"¨(⊂[2]M)COMPRESS¨⊂ι1↓ρM

To generalize this phrase, further account should be taken of the possibility of all-zero rows. Empty vectors must be eliminated from the result of **COMPRESS**¨ and matching row indices from the left argument of ,"¨ which leads to

        ((∨/M)/ι↑ρM),"¨((⊂[2]M)COMPRESS¨⊂ι1↓ρM)~⊂ι0

Illustration : Binary matrix as partitions of column indices

The first row of the matrix M in the previous illustration :

        M41
  1 1 0 0
  0 0 0 1
  1 0 0 0

partitions ι4 into 3 4 corresponding to 0s and 1 2 corresponding to 1s, the second row partitions ι4 into 1 2 3 and 4, and the third row partitions ι4 into 2 3 4 and 1. The necessary masks can be described by

        (~M41)M41
  0 0 1 1   1 1 0 0
  1 1 1 0   0 0 0 1
  0 1 1 1   1 0 0 0

and translated into indices by applying **COMPRESS**¨ with a further level of nesting in the left argument

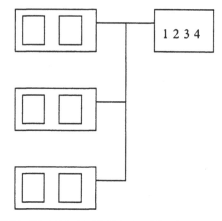

DISPLAY(⊂[2]¨(~M41)M41)COMPRESS¨¨⊂⊂ι4

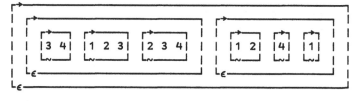

Exercises 4b

1. Assume DTB is defined as in Exercise 2c6 to delete trailing blanks from a character vector.

a. Use this function to write another function DBTM which converts a character name matrix (i.e. one in which each row is a name, and the shorter names are padded on the right with blanks) into a vector of names, each with no trailing blanks.

b. Write a function Z←L INDEX R for which R is a simple character string, L is a name matrix, i.e. a matrix each of whose rows is a name, possibly padded with blanks, and Z is the indices of all rows of the matrix which contain R, if necessary padded with blanks.

2. a. Write an expression which makes a character matrix consisting entirely of digits and spaces into a numeric matrix possibly padded with zeros.

b. Generalize your expression to deal with arrays of any dimension.

4.3 Distinctions between Similar Primitives

APL2 has several alternatives for selecting items from a nested array. The group of functions

first, take, pick, disclose and **index**

are discussed together since they have both semantic and graphical affinities with each other. In the context of this discussion, index applies equally to "squad" indexing and to bracket indexing.

4.3.1 First and Take/Drop

A significant attribute of functions is their effect with regard to depth. **First** is depth-reducing, **take** and **drop** are not.

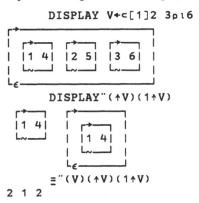

```
        DISPLAY V←⊂[1]2 3ρι6
```

```
        DISPLAY¨(↑V)(1↑V)
```

```
        ≡¨(V)(↑V)(1↑V)
2 1 2
```

Thus ↑ and 1↑ offer a straightforward choice between penetration (i.e. depth-reduction) and cross-sectioning (cf. Section 4.1 Pick and Indexing). For example ↑ι8 is scalar 1, 1↑ι8 is vector 1. Further, ↑ will do a **ravel** if necessary and is thus guaranteed not to give LENGTH ERRORs such as arise from, say

```
        1↑2 2ρι4
```

For all non-empty arrays A the following identity holds

```
        (↑A) ←→ (⊂T)⊃(T←(ρρA)ρ1)↑A
```

where (ρρA)ρ1 should be considered as a *path*. The presence of the **pick** on the right hand side emphasizes the need to reduce depth by one in order to relate a **first** to a **take**.

Take and **drop** are the subject of two identities:

```
        (ρI↑A) ←→ |I
        (ρI↓A) ←→ 0⌈(ρA)-I
```

where I is an index vector. These formalize the rank-preserving property of **take** and **drop**.

Idioms involving **first** are more numerous than those involving **take**, for example

```
A←2 5 2 3ρι60
```

```
        ↑A        ⍝ first item of a simple array
1
        ↑⌽,A      ⍝ last item of a simple array
60
        ↑ρA       ⍝ size of first dimension
2
        ↑⌽ρA      ⍝ size of last dimension
3
```

Combining **first** and **drop** leads to some useful phrases, e.g. ↑I↓V selects the Ith item in vector V regardless of the setting of ⎕IO. Hence

```
        ⍢↑condition↓⌽ 'then clause'  'else clause'
```

implements if-then-else (see Section 5.6.2 for another way of doing so), or more generally

```
        ⍢↑I↓  ''  'case1'   'case2'   ...   'casen'
```

implements a case statement, for example

```
        COND2←1+COND1←1+COND0←0
        ⍢↑COND0↓'0' '10' '20'
0
        ⍢↑COND1↓'0' '10' '20'
10
        ⍢↑COND2↓'0' '10' '20'
20
```

↑¨ *penetrates* but does not *remove* the outer level of structure. Instead it removes the levels below the outer one so that the identity

```
        ρ↑¨A  ←→  ρA
```

holds. For example

```
        DISPLAY V←⊂[1]2 3ρι6
```

```
        ≡↑¨V
1
```

The rank of each item of an array is given by:

```
        ↑¨ρ¨ρ¨A
```

For example:

```
M42←2 2ρ(2 2ρ'ABC'(2 2ρ'X'))6(1 3ρ5)(ι3)
DISPLAY M42
```

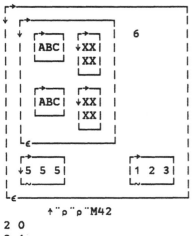

```
            ↑"ρ"ρ"M42
2 0
2 1
```

Another phrase involving ↑ strips off one level of depth from within the outermost layer:

```
↑,/,¨V
```

for example

```
V44←((2 2ρ'ABC')(2(3 4))(5 6))
DISPLAY V44
```

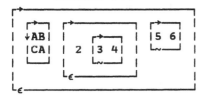

```
DISPLAY ,¨V44
```

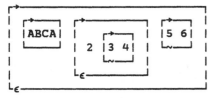

```
      DISPLAY ,/,¨V44
```

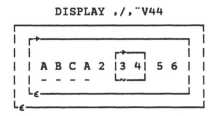

(see Section 5.5.2 for a detailed description of `,/`.)

```
      DISPLAY ↑,/,¨V44
```

```
┌→──────────────────────────────┐
│                ┌→──┐            │
│ A B C A 2 │3 4│ 5 6 │
│ - - - -   └~──┘            │
└∊───────────────────────────────┘
```

```
      ρ↑,/,¨V44
8
```

Contrast this with ∊ which strips **all** levels of nesting:

```
      DISPLAY ∊V44
```

```
┌→───────────────┐
│ABCA 2 3 4 5 6│
└+───────────────┘
```

```
      ρ∊V44
9
```

4.3.2 First and Pick

The depth of the right argument of dyadic `⊃` (**pick**) is reduced by the number of items in its left argument or **path**.

```
      V←⊂[1]2 3ρι6
```

```
      DISPLAY¨(V)(2⊃V)(2 1⊃V)
```

```
┌→──────────────────────┐  ┌→──┐
│ ┌→──┐ ┌→──┐ ┌→──┐ │  │2 5│    2
│ │1 4│ │2 5│ │3 6│ │  └~──┘
│ └~──┘ └~──┘ └~──┘ │
└∊──────────────────────┘
```

```
      ≡¨(V)(2⊃V)(2 1⊃V)
2 1 0
```

The above rule extends to the case of an empty path, in which case the result is simply the right argument, that is `ι0` is the left identity of **pick**.

```
      (ι0)⊃V
1 4  2 5  3 6
```

`1⊃V` is equivalent to `↑V` for any **non-empty** vector **V**. While both are depth-reducing, the former is valid only if **V** has at least one item, otherwise an INDEX ERROR occurs. **First** (`↑`) by contrast never returns an INDEX ERROR. If there is no first item, **first** supplies a *fill item*. In the first two examples below the fill item is a scalar, in the third example it is an empty vector.

        DISPLAY ↑0ρ2 3ρ5

0

        DISPLAY ↑0ρ2 3ρ'ABC'

–

        DISPLAY ↑⊂ι0
┌⊖┐
│0│
└~┘

4.3.2.1 Type and Prototype

The value of the fill item arising from applying ↑ to an empty array depends on how the empty array was constructed in the first place. Although empty arrays are objects with structure but without data, they have to be created out of *some* data initially, and the originating data is reflected in the value of ↑A. For example:

        A←3 4 0ρ(⊂2ρ9)(⊂3ρ9)
        B←3 4 0ρ(⊂3ρ9)(⊂2ρ9)

        ↑A
0 0
        ↑B
0 0 0

The above results are called the *prototypes*, which means literally the types of the **first**s, of A and B. Only the leading item from which an array is created influences its prototype, e.g. the fact that the second item from which A above is created has shape 3 is not reflected in the prototype.

The prototype of an empty array A is thus a *non-empty* array and its data does not reproduce the data which was used in constructing A in the first place, which cannot be recaptured by further processing. For example

        1 2 2↑↑A
0 0 0 0
0 0 0 0

contains no reference to 9, although the type and structure of A are inherited by the prototype.

It is often desirable to construct an array of identical structure to a general array A with 0 replacing numbers, blank replacing characters, and ι0s remaining unchanged. This is achieved by enveloping A in a further level of nesting and then deliberately constructing an empty array 0ρ⊂A whose prototype has the desired property. The result of the expression ↑0ρ⊂A is called the *type* of A. In passing note the approximate graphical similarity between the word "type" and the APL character string "↑0ρ⊂." Here is an example:

```
V45←(137 'ABCDEF'(45 'G'))
DISPLAY V45
```

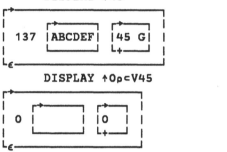

```
DISPLAY ↑0ρ⊂V45
```

Illustration : Distinguishing character, numeric, etc.

Defining

```
[0]    Z←TYPE R
[1]    Z←↑0ρ⊂R
```

the expression 2⊥0 ' '∊∊TYPE A returns 0 for empty, 1 for character, 2 for numeric, 3 for mixed.

The **first** of the **ravel** of the non-empty array TYPE A can now be defined as the prototype of A, thus extending the definition of prototype to non-empty arrays.

Prototypes are most useful when dealing with objects which are uniform in structure, since in this case the structure of the leading item reflects the construction of all items. They are also used in situations where "proxy" data is needed by functions which either

(1) enlarge structure (but not data) either at the extremities (**take**) or at an arbitrary point within the structure (**expand** and **replicate**); or

(2) pad structure to rectangularity (**disclose**).

The fill item is different in the two cases above, viz. in case (1) the prototype of the entire array is used, e.g.

```
V46←((1 2)3)(4 5 6)(7 8)
DISPLAY 4↑V46
```

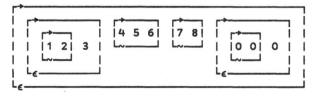

DISPLAY 1 0 1 1\V46

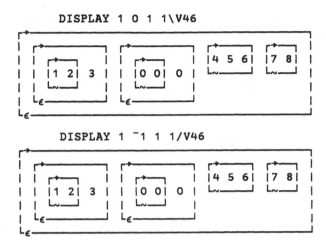

DISPLAY 1 ¯1 1 1/V46

In case (2) the prototypes of *cross-sections* of the array are used, that is ↑¨0ρ¨V:

```
     ⊃V46
 1 2   3   0 0
   4   5   6
   7   8   0
```

IBM manuals have defined prototype as ↑0ρ⊂↑, that is as **TYPE first**, however two of the characters in this phrase are redundant in that

```
     ↑0ρ⊂↑A   ←→   ↑0ρA
```

Informally ⊂ (**enclose**) replaces the depth that ↑ (**first**) removes. Further if A is empty, or even if only ↑A is empty, then the prototype is ↑A.

To summarize, prototype is ↑0ρA, reducing to ↑A if ↑A is empty.

4.3.3 Pick and Disclose

Both **disclose** (⊃) and **pick** (↑) reduce the depth of an array, however **pick** is selective as well. In selecting an item **pick** may penetrate several levels of structure. **Pick** always returns the selected item as it was nested in the structure. **Disclose** may result in padding with fill items. **Disclose** returns all of the original data but it is transformed into an object with one less level of depth. It changes the *structure* of the original object, but requires uniformity of rank one level down, that is all items of ρ¨ρ¨A must be identical, subject to some flexibility on account of scalar extension. Thus the following is an error situation:

```
     A←3 2ρι6
     V←'ABC'
     VA←V A
     ⊃VA
RANK ERROR
     ⊃VA
     ∧∧
```

If rank uniformity one level down *does* apply, the shape of ⊃A is the shape of A with ⌈/ρ¨A catenated to its right. Formally

        (ρ⊃A) ↔ (ρA),⌈/¨ρ¨A

When axis qualification is used with **disclose** used, it indicates where the nested shape values are to appear in the disclosed array. Here are some examples:

        ρ(1 2)(3 4 5)
2
        ρ⊃(1 2)(3 4 5)   ⍝ default axis for ⊃ is 2
2 3   ⍝ 3 comes from inner structure to second
                                      position of new structure
        ⊃(1 2)(3 4 5)
1 2 0
3 4 5
        ρ⊃[1](1 2)(3 4 5)
3 2   ⍝ 3 comes from inner structure to first
                                      position of new structure
        ⊃[1](1 2)(3 4 5)
1 3
2 4
0 5

The next example demonstrates the effect of using a vector as an axis qualifier:

        A←3 2ρι6
        B←2 4ρ10×ι8
        AB←A B
        ρAB
2
        ρ¨AB
3 2   2 4
        DAB←⊃AB
        ρDAB
2 3 4
        ρ⊃[1 2]AB
3 4 2
        ρ⊃[1 3]AB
3 2 4

When V is a vector of vectors ⊃V is often useful for displaying V as a matrix, thereby make the relation between corresponding items clearer.

Illustration : Converting vector of names to a matrix form

While some functions such as **grade-up** require simple matrix arguments, names are often more easily entered as a vector of vectors, e.g.

        NAMEVV←'NORMAN' 'JEAN' 'JO' 'RAY'

Disclose gets them into a matrix form:

```
      MNAMES←⊃NAMEVV
      ρMNAMES
4  6
      MNAMES
NORMAN
JEAN
JO
RAY
```

4.3.4 First and Disclose

Suppose

```
      A←⊂?2 2ρ10
```

In this instance ⊃A and ↑A have identical results but the routes by which they are reached are quite different. The effect of **disclose**, ⊃, is to **restructure** the entire nested object by bringing the shape vector 2 2 from the inner structure to the outer structure.

By contrast **first** (↑) does a depth-reducing *selection* which penetrates one level of depth and selects the first object it finds there, in the case of A the 2x2 matrix within the nested scalar.

Both functions are inverse to **enclose**, that is ↑⊂A and ⊃⊂A are both equivalent to A. **Enclose** however is *not* the inverse of either, i.e. in general neither ⊂↑A nor ⊂⊃A is equivalent to A.

4.3.5 Summary of Relationship between above Functions

It is useful to summarize the relationships between these functions both in the form of a table:

		Depth- reducing?	Fill/ Index-err?	Structural/ Selection?
⊃	Disclose	Y	F	STR
↑	Take	N	F	SEL
↑	First	Y	F	SEL
⊃	Pick	Y	I	SEL
⎕	Index	N	I	SEL

and also in the form of a diagram in which a double line indicates that the two functions at its ends share two attributes from the table, a single line that they share just one.

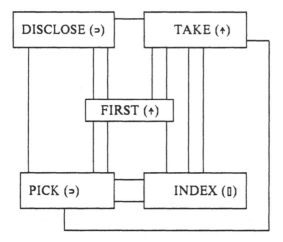

Both the table and the diagram show how **first** has the the greatest overall affinity with the others in the set. Also although **pick** and **disclose** share a common symbol they have only one of the three attributes in common.

Exercises 4c

1. Given

```
A+2 3ρι6      B+3      C+'APL'      E+(⊂A),B,⊂C
```

what are value, shape and depth for each of the following :

a. OρE   f.   ↑E[2]
b. ↑OρE   g.   ⊃E[2]
c. ↑Oρ⊂E   h.   ↑OρE[2]
d. ↑Oρ⁻1↑E   i.   ⊃OρE[2]
e. ⊃Oρ⊃E   ?

2. For any array A describe fully the differences between

   a. ↑A   b. 1↑A   c. 1⊃A   d. A[1].   e. 1⎕A

Which, if any, are the same if

(i) A+(1 2)(3 4)   (ii) A+ι8

3. Define D↔(1 2)3'ABC'.

   a. What is (i) the type; (ii) the prototype of D ?

   b. What, if any, is the difference between the prototypes of D and φD ?

   c. What are (i) 5↑D (ii) 5↑¨D (iii) 5↑φD ?

 d. What are (i) ⊃D (ii) ⊃¨D (iii) ⊃[1]D ?

4. a. What is ⊃'THIS' 'IS' 100 ?

 b. Obtain the matrix below using ⊃

```
1 0 0
1 2 0
1 2 3
```

 c. Transform   V47←('JACK' 10)('PETER' 27)   into

```
JACK PETER
   10    27
```

using only ⊂ ⊃ and axis qualification.

5. If V is a non-empty vector, explain why the following two expressions which both extract the last item do not match:

 a. ↑⌽V   b.   ¯1↑V

6. If M is 2 3ρι6 what is the difference between

 a. ↑[1]M   b. 1↑[1]M   c. ↑⊂[1]M   d. ↑¨⊂[1]M   ?

7. If M is a matrix, write the following expressions more briefly:

 a. 2 1⎕¨⊂ρM   b.   (⊂2 1)⎕ρM

8. Make a name list (that is a vector of character vectors) of the names appearing below in ascending order of wealth:

 V48←('TRUMP' 8.1)('GETTY' 7.4)

9. If A is an array, what single primitive function is equivalent to

 (⊂T)⊃(T←(ρρA)ρ1)↑A   ?

4.4 Empty Arrays and Fill Functions

A language which admits empty arrays must address the problem of what each primitive function should do when presented with an empty array, that is an object with structure but no data. In such cases function execution in the ordinary sense of transforming data is not applicable and consequently each APL2 interpreter must contain rules for providing the necessary padding characters. These rules are embodied in the so-called *fill functions*. Historically these have not been consistent across different APL2 interpreters. For scalar primitive functions F the result of F EA (EA = Empty Array) or EA1 F EA2 reflects a combination of prototype, shape and fill function subject to rank and length conformance. Here are some examples on IBM mainframe APL2:

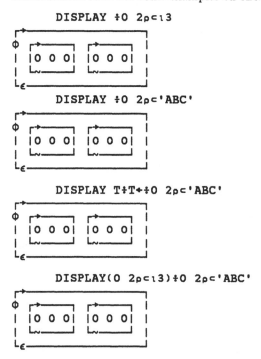

In APL2 the fill function of all scalar primitives with scalar arguments is a function whose result is always zero even if the type of R is character. For non-simple arguments the fill function is applied to each item recursively until simple scalars are reached - this is consistent with the pervasive property of scalar primitive functions in preserving shape and depth in the absence of data. When one argument is scalar but non-empty and the other is empty, e.g. for

        2+0ρ⊂0 0

the fill function is applied to the prototype of the non-empty argument as one argument and the empty argument as the other, so that the result of the above expression is

```
      DISPLAY 2+0ρ⊂0 0
```

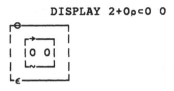

The monadic and dyadic ⊞ functions are a special case since they are arithmetic but non-pervasive, and so what happens when **each** to them is applied is not dealt with by pervasion. Since F¨ is equivalent to defining a loop for F, it is desirable to maintain consistency of shape even when the loop is executed zero times, which is what happens when F¨ is applied to an empty array. The fill functions in APL2 for monadic and dyadic ⊞ are:

```
[0]    Z←MFF R            [0]    Z←L DFF R
[1]    Z←⍉↑R              [1]    Z←((1↓ρ↑R),1↓ρ↑L)ρ⍉R
```

For example:

```
      ↑⊞¨0ρ⊂2 3ρ0
0 0
0 0
0 0

      B←?3 2ρ10
      C←?3 3ρ100
      B⊞C
 0.06595 ⁻0.02702
⁻0.05348  0.1279
 0.04424 ⁻0.03799
      DISPLAY (0ρ⊂B)⊞¨0ρ⊂C
```

Non-scalar primitive functions have their results defined at the structure phase and so no separate fill functions are required, e.g. the results of

```
      DISPLAY ρ⊂0 3ρ0
```

```
      DISPLAY ρ⊂4 5ρ'ABCDE'
```

are not affected by the emptiness or otherwise of whatever is to the right of ⊂ nor by its type, and likewise

```
      'ABC'⍳⊂0 3ρ0
```

depends only on whether a match is found for the enclosed scalar, and not on whether or not it is empty.

4.4.1 Identity Items and Identity Functions

There is an important distinction between the concept of identity *items* of scalar dyadic functions which satisfy

```
A F IDI  ↔  A          (Right identity item)
IDI F A  ↔  A          (Left identity item)
```

and identity *functions* which satisfy

```
L F (IDF L)  ↔  L      (Right identity function)
(IDF R) F R  ↔  R      (Left identity function)
```

4.4.1.1 Identity Items

A full set of identity *items* is given in the following table

+	-	×	÷	|	⌊	⌈	*	⍟	○	!	∧	∨	<	≤	=	≥	>	≠	⍲	⍱
0	0	1	1	0	M	-M	1		none	1	1	0	0	1	1	1	0	0	none	
LR	R	LR	R	L	LR	LR	R			L	LR	LR	L	L	LR	R	R	R		

where L and R stand for left and right and M is the largest number in the machine. These items are obtainable as F/⍳0. The reason is that

```
(F/V) F (F/V1)  ↔  F/V,V1
```

is a fundamental property of pervasive functions and so setting V1 to ⍳0 gives

```
(F/V) F (F/⍳0)  ↔  F/V
```

4.4.1.2 Identity and Inverse Functions

Fulfilling the left-identity-function identity above for a function such as ρ requires a data-dependent argument, viz.

```
(ρR)ρR  ↔  R
```

that is ρ is its own left identity function. Now take the function ⍴. What left argument makes

```
L⍴R  ↔  R  ?
```

The answer is ⍳ρρR so ⍳ρρ is the identity function. A full list of identity functions is

Fn.	Identity function	L/R	Restriction
ρ	ρR	L	
,	$((^{-}1↓ρR),0)ρ⊂((^{-}1↓ρR),0)ρR$	LR	1≤ρρR
φ	$(^{-}1↓ρR)ρ0$	L	
⊖	$(1↓ρR)ρ0$	L	
⍉	ιρρR	L	
⊃	ι0	L	
↓	$(ρρR)ρ0$	L	
↑	ρR	L	
~	ι0	R	1=ρρL
⌷	$(ι↑ρR)∘.=ι↑ρR$	R	1≤ρρL

Identity functions should not be confused with *inverse* functions which are defined by

```
L LIF L F R    ↔    R        (Left inverse function)
(L F R) RIF R  ↔    L        (Right inverse function)
```

The only scalar dyadic functions which possess inverse functions are

	+	−	×	÷	⋆	⍟	=	≠
Left		−		÷	⍟	⋆		=	≠
Right	−		+		×			=	≠

Exercises 4d

1. What are

a. ×/2 3 0ρ0
b. ×/2 0 3ρ0
c. ×/⊂2 0 3ρ0 ?

What difference does it make if the rightmost 0s are replaced with 9s?

2. What are

a. ↑(ι0)÷0ρ⊂2 3ρ0
b. ↑⌷¨0ρ⊂2 3ρ0
c. ↑(⊂3 4)ρ¨0ρ⊂2 3ρ0 ?

3. What are

a. ρ/2 3 0ρ0
b. ρ/0ρ⊂2 3ρ0
c. ρ/0ρ⊂Vρ0   where V is any simple numeric vector ?

4. What is the value of `F/0ρ⊂2 3ρ0` when `F` is:

a. ⍉   b. ⊃   c. ↑   d. ↓   e. ~   f. ,   ?

5. What are

a. `ρ↑,/0ρ⊂2 9ρ0`
b. `ρ,/0ρ⊂2 9 9ρ0`
c. `ρ↑,/0ρ⊂2 9 9 9ρ0` ?

What happens eventually as more 9s are added?

Summary of Functions used in Chapter 4

Section 4.2
SENTENCE   builds a random sentence

Section 4.2.1
VCAT   catenates matrices vertically
PENCL   descalarization with partial enclose
ONES   obtains co-ordinates of 1s in a binary matrix

Exercises 4b
DTBM   deletes trailing blanks in each row of a matrix
INDEX   indices of rows in matrix containing given string

Section 4.3.2.1
TYPE   type of an APL array

5
Using Operators

5.1 The Role of Operators in APL2

Although nested arrays are the most distinctive feature of APL2, operator extensions provide at least as great an advance. There are two aspects to operator extension - first the provision of user-defined operators, and secondly the extension of existing operators to nested arrays and to user-defined functions and derived functions. These two features increase by a huge factor the expressiveness of APL2 in describing programming ideas.

In general if APL objects are the "nouns" of the language and functions the "verbs," then operators are the "adverbs." They direct *how* to apply or combine functions in ways which are common across a range of functions. From the earliest days of APL the adverbial aspect of functions was achieved by "embroidering" the function symbols with other symbols such as / and .. For example, +/V describes *how* to add the elements of the vector V, i.e. add "through" V, in the sense of inserting the function + into all the available spaces (one less than the number of elements in V) and evaluating the resulting expression. The insertion is a *structural* action and the consequent evaluation a *functional* one. After defining "reduce" to describe the adverbial concept of "through," it makes sense to talk about "multiply reduce," "divide reduce," and so on.

5.1.2 User-defined Operators

The mechanics of creating a defined operator are similar to those for defining a function. Four independent choices are made, viz:

its name

the number of arguments

the number of operands

whether or not it has explicit result

As a general principle a user-defined operator should be constructed when several functions are handled in the same way. A simple example is COM which reverses or "commutes" the role of the left and right arguments.

```
[0]    Z←L(P COM)R
[1]    Z←R P L

       2 -COM 5
3
```

An operator has either one or two operands which are denoted here by P (left operand) and Q (right operand - if present). P COM is the derived function and L and R are its left and right arguments. The binding of the operator COM with the function **minus** is stronger than that between the derived function and the two arguments, and so the above expression should be read in the three "chunks" which are suggested by the spaces.

Tracing function execution is another situation in which different dyadic functions are handled in the same way. The operator in this case is called SEE.

```
[0]    Z←L(P SEE)R
[1]    L 'f' R '=' Z←L P R
```

-SEE is the same as - except that an explicit message is issued for every **minus** execution.

```
        -SEE/ι4
3 f 4 = ‾1
2 f ‾1 = 3
1 f 3 = ‾2
‾2
```

Operators can be be used in conjunction, e.g.

```
        -SEE COM/ι4
4 f 3 = 1
1 f 2 = ‾1
‾1 f 1 = ‾2
‾2
```

Since -SEE is the same as **minus** except for messages it follows that -SEE COM is the same as -COM except for messages. The order of operators is important - in the above traces f denotes - whereas in the next sequence f denotes -COM.

```
        -COM SEE/ι4
3 f 4 = 1
2 f 1 = ‾1
1 f ‾1 = ‾2
‾2
```

Illustration : Moving functions along an axis

Reduction applied to the non-commutative functions **subtract** and **divide** produces alternating sums and products respectively. This is counter-intuitive when viewed with eyes conditioned by ordinary arithmetic in that `-/6 2 3` *looks* as though it means `6-2-3`, i.e. `1` rather than `7`. The latter is indeed a reasonable variant of subtraction - call it subtraction "along" a vector - and the following operator describes recursively the process for a general scalar dyadic function P, which could be either primitive or user-defined:

```
[0]   Z+(P ALONG)R
[1]   +L1 IF 1=ρ,R            ⍝ branch if singleton or scalar
[2]   +0 Z+((P ALONG)¯1↓R)P↑ΦR  ⍝ apply P once at right hand end of R
[3] L1:Z+↑R
```

```
      (-ALONG)ι4
¯8
      ÷(÷ALONG)ι4
24
```

This operator can be further generalized by specifying an axis Q as a second operand.

```
[0]   Z+(P Along Q)R
[1]   +L1 IF 1=Q⎕ρR
[2]   +0 Z+((P Along Q)¯1↓[Q]R)P 1↑[Q]ΦR
[3] L1:Z+1↑[Q]R
```

```
      (-Along 2)2 3ρι6
¯4
¯7
```

The result of either function **ALONG** or **Along**, unlike that of reduce, has in general the same rank as its argument.

The second program lines of the functions in the above illustration exhibit a technique which is widely used in the remainder of this book, particularly in recursive functions. It consists of using the characters `+0 Z+...` to compress assignment and branching into a single line. The effect of this is to make an intermediate nested object of depth one greater than the object of the assignment, and then force an implicit **first** for the branch.

Another straightforward example of a simple operator is the outer product of a
vector **V** with itself.

```
[0]    Z+(P TABLE)R
[1]    Z+R•.P R
```

```
       ×TABLE ι12
```

is thus the ordinary school multiplication table.

A common property of the operators **ALONG,** **TABLE**, and **COM** is that they
can be applied across a *set* of functions. If an operator were relevant to only one
function, it would be preferable to write a user-defined function.

Of the operands P and Q, either or both may be functions or arrays, but
usually at least one is a function. The derived function then takes arguments L
and R. If Q and R are both arrays as in the case of the function **ALONG** then
parentheses may be necessary to show where Q stops and R begins.

A monadic function has only a *right* argument. A monadic operator on the
other hand has only a *left* operand. Non-ambiguity of syntax demands that
operators follow the opposite rule to that for functions, that is they are executed
from left to right. Thus in the expression

```
       -COM ALONG 1,ι4
2
```

the operator **COM** is executed before the operator **ALONG**. More specifically the
derived function -**COM** is constructed and then passed as an operand to **ALONG**
to obtain the derived function (-**COM**) **ALONG** 1. Applying this derived function to the right argument, the successive execution steps are :

```
       1(-COM)2   =   1
       1(-COM)3   =   2
       2(-COM)4   =   2
```

giving 2 as the final answer. The above expression also raises the issue of where
the right operand stops and the right argument begins, and under what conditions explicit parentheses are necessary. The precise rules for determining such
matters are called the *binding rules* and they are discussed in the next Section.

5.2 Binding

For expressions which contain only functions and variables the evaluation rule
known as the "right-to-left evaluation" rule applies, viz:

The rightmost function whose arguments are available is evaluated first.

Including an operator in an expression requires further rules. The expression

        -COM ALONG 1,ι4

discussed in the previous Section raised the issue of whether 1 is a right operand
to **ALONG** or the left argument to the function **catenate**. At first glance the
comma appears to represent the catenation 1,ι4 . However 1 is not a left
argument to catenate because it has a higher priority as right operand to the
operator **ALONG**. The 1 *binds* more strongly to the operator as an operand than
it does as an argument to a function so that the expression is equivalent to

        (-COM ALONG 1) ι4

This example demonstrates that in addition to the right-to-left evaluation rule, a
set of *binding strengths* between operators, functions and other syntactic symbols
needs to be defined. Binding rules define how variables and symbols group for
evaluation. For any three objects A B C the following binding table determines
whether B associates with A or C, that is whether A B C means (A B) C or A
(B C).

Binding Strength	Object	Binds to-
(Strongest)	1. Bracket	the item on its left
	2. Assignment	the name on its left
	3. Operator	its right operand
	4. Vector item	the items on either side
	5. Operator	its left operand
	6. Function	its left argument
	7. Function	its right argument
(Weakest)	8. Assignment	whatever is to the right

As an example of how to use the table consider the problem of deciding whether
the expression +.×/A means (+.×)/A or +.(×/)A. The binding between .
and × (right operand binding) is stronger than that between × and / (left
operand binding) and so the inner product is evaluated *before* the reduction.
The entire expression thus means (+.×)/A and *not* +.(×/)A. The application
of the binding rules to expressions containing two or more operators can be
expressed more generally as

*Operators have long left scope and short right scope whereas functions have
long right scope and short left scope.*

Illustration : Implications of Binding

a. Consider

```
X←10
5+X←2
```
7

The binding of X to ← in the second line is stronger than that to +, otherwise the value of the result would have been 15.

b. The expression

```
2 4 6[2]
```

results in a **RANK ERROR** since the bracket binds more strongly to 6 than does 6 to the vector 2 4. To achieve what is presumably the desired indexing the binding strengths must be overruled with parentheses:

```
(2 4 6)[2]
```

c. Define

```
[0]   Z←(P RED Q)R
[1]   Z←P/[Q]R

      +RED 2 3 2ρι6
3 7 11
```

is equivalent to

```
      (+RED 2)3 2ρι6
3 7 11
```

since the binding of RED to its right operand 2 is stronger than the binding of 2 as an item of the vector 2 3 2.

Operators may include other operators in their definition, e.g. reduction from the left is given by

```
[0]   Z←(P LRED Q)R
[1]   Z←P COM/[Q]⌽R

      (-LRED 2)2 3ρι6
⁻4 ⁻7
```

LRED is similar to **ALONG**, but mimics reduction more closely than **ALONG** by reducing rank. Although functions derived from user-defined operators may be ambi-valent (see Section 5.3.2), operators themselves are not, so an attempt to use LRED monadically results in e.g.:

```
        (-LRED)2 3ρι6
SYNTAX ERROR
        (-LRED)2 3ρι6
        ∧       ∧
```

Illustration : Hexadecimal Arithmetic

The operator HEX transforms arithmetic functions into their equivalents for performing hexadecimal arithmetic.

```
[0]     Z←L(P HEX)R           ⍝ L and R are character strings
[1]     Z←DTH(HTD L)P HTD R   ⍝   representing hex numbers
```

The functions DTH and HTD convert decimal to hex and hex to decimal respectively.

```
[0]     Z←HTD R               ⍝ R is a hex character vector
[1]     Z←16⊥¯1+HSTRιR
```

```
[0]     ∇Z←DTH R              ⍝ R is a numeric array
[1]     Z←HSTR[1+((⌊1+16⍟1⌈⌈/,R)ρ16)⊤R]
```

where HSTR is the character string '0123456789ABCDEF'. Here are some examples:

```
        'A1'+HEX'4F'
FO
        'A1'×HEX'4F'
31AF

        +HEX/'F3' '8' '2'
  FD
        '12'+HEX¨'F3' '8'       ⍝ 1 added to X'F3', 2 added to 8
  F4 A
        (⊂'12')+HEX¨'F3' '8'    ⍝ X'12' added to both X'F3' and 8
  105 1A
```

Comparison of the last two examples shows how **enclose** is necessary in order to have '12' interpreted as a single hex integer, with subsequent scalar expansion (see Section 2.1.2). This suggests that the operator HEX be extended to HEXE (standing for **HEX each**) thus:

```
[0]     Z←L(P HEXE)R
[1]     Z←DTH¨(HTD¨L)P HTD¨R

        'A1' '12'+HEXE '4F' 'F3'
 FO 105
        (⊂'12')+HEXE 'F3' '8'
 105 1A
```

Illustration : An Operator for Padding Matrix Catenations

This illustration shows how variations on text-joining with regard to axis and justification can conveniently be brought together by defining an operator. The technique can be compared with **VCAT** given in Section 4.2.1.

```
[0]    Z+L(P NEXT Q)R;T;U
[1]    (L R)+MATRIFY"L R        ⍝ ensure both arguments are matrices
[2]    Z+Qx(ρL)⌈ρR              ⍝ Z is used as a local variable ...
[3]    T+Z+(ρL)xU+~|Q           ⍝ ... to calculate left arguments for take ...
[4]    U+Z+UxρR                 ⍝ ... prior to catenation
[5]    Z+(T↑L),[P]U↑R
```

```
[0]    Z+MATRIFY R
[1]    Z+(⁻2↑1 1,ρR)ρR
```

Neither operand of **NEXT** is a function, so effectively **NEXT** is a function with four arguments. The left operand P is the axis qualifier for catenation and the right operand Q is a code which determines in which direction (if any) to apply padding in order to make the smaller dimension match the larger. Its domain is ⁻1 0 1; 1 means pad ↓ or →, ⁻1 means pad ↑ or ←, 0 means don't pad. A code of 0 in the Pth. item of Q can give rise to potential **LENGTH ERROR**s. A few examples should make the operation clear.

```
          DISPLAY"M51 M52
┌→─────┐   ┌→───┐
↓BREAD│   ↓MAN│
│FRUIT│   │CAN│
└─────┘   │EAT│
          └───┘
```

```
          DISPLAY"(M51(1 NEXT(0 1))M52) (M51(1 NEXT(0 ⁻1))M52)
┌→─────┐   ┌→─────┐
↓BREAD│   ↓BREAD│
│FRUIT│   │FRUIT│
│MAN  │   │ MAN│
│CAN  │   │ CAN│
│EAT  │   │ EAT│
└─────┘   └─────┘
```

```
          DISPLAY"(M51(2 NEXT(1 ⁻1))M52) (M51(2 NEXT(⁻1 0))M52)
┌→────────────┐   ┌→─────────┐
↓BREAD   MAN│   ↓     MAN│
│FRUIT   CAN│   │BREADCAN│
│        EAT│   │FRUITEAT│
└───────────┘   └─────────┘
```

```
          DISPLAY"(M51(2 NEXT(1 0))M52) (M51(2 NEXT(⁻1 ⁻1))M52)
┌→────────┐   ┌→─────────┐
↓BREADMAN│   ↓     MAN│
│FRUITCAN│   │BREAD CAN│
│     EAT│   │FRUIT EAT│
└────────┘   └─────────┘
```

Exercises 5a

1. The expression

```
T/ιρT←((N-1)ρ1)≤2</V
```

in which V is a simple numeric vector and N a positive integer returns the indices in V of the starting points of all strictly increasing subsequences of length N, e.g.

```
V←2 3 4 3 4 5 2 2 7
N←2
T/ιρT←((N-1)ρ1)≤2</V
1 2 4 5 8
```

Write an operator CONSEC which allows the determination of the equivalent information for

   a. strictly increasing sequences;
   b. strictly decreasing sequences;
   c. non-decreasing sequences;
   d. non-increasing sequences;
   e. sequences of equal values;
   f. sequences of values in which every item differs from its neighbor.

2. a. Write an operator BASE which performs arithmetic functions P on scalar numeric integers which are to be interpreted as integers in base Q, e.g.

```
16+BASE 7 23
42
1111÷BASE 2 11
101
```

   b. How would you extend this to process integer arrays, so that you could for example divide the two by two array

```
 1111     110
10010 100001
```

by 11 in base 2.

3. Describe the difference between

```
[0]    Z←L ROOT R
[1]    Z←R*÷L
```

and

```
[0]    Z←(P ROOTOP)R
[1]    Z←R*÷P
```

In what circumstance might it be desirable to use ROOTOP rather than ROOT?

4. With COM and SEE as defined above, which if any of the following expressions are *necessarily* identical for a general numeric vector V?

  a. -SEE/V   b. -SEE COM/V   c. -COM SEE/V

5.3 Matching Function Arguments

5.3.1 Function Composition

Composition means the successive application of two functions. For example consider

        V12←12(13(14 15))(16 17)

The composition (ρε) applied to V means apply ε to V followed by ρ, and so is rendered by

        ρεV12
6

The **each** of this composition is given by first applying ε to each item of V:

        ε¨V12
 12   13 14 15   16 17

and then ρ¨ to the result:

        ρ¨ε¨V12
  1   3   2

Similarly the **each** of the composition (+/ε) is given by

        +/¨ε¨V12
12 48 33

The general rule for applying **each** to compositions of monadic functions is:

        (PQ)¨ ←→ P¨Q¨

For dyadic function compositions a left argument has to be allocated to one of the two component functions. Sometimes only one allocation of the left argument is sensible. For example consider the composition ⍋⍉:

        'ABC'⍋⍉'CAT'
 2 3 1
        'XYZ'⍋⍉'AXE'
 2 1 3

2 3 1 and 2 1 3 are the **grade-up** vectors with the alphabets 'ABC' and 'XYZ' respectively of the words 'TAC' and 'EXA'. In this case the rule stated above may be applied, that is (⍋⍉)¨ is given by

        'ABC' 'XYZ'⍋¨⍉¨'CAT' 'AXE'
  2 3 1   2 1 3

and the left argument applies to the leftmost of the two functions in the composition.

Where the functions P and Q both have possible dyadic meanings, ambiguity can arise as to whether a left argument applies to P or Q. For example if the composition $\epsilon\iota$ is applied with left argument $(\iota3)$ and right argument of 2 the left argument can relate to either the ϵ or the ι, thus:

```
      (ι3)ει2
1 1 0
      ε(ι3)ι2
2
```

In such cases composition must be defined explicitly via a defined operator. For example:

```
[0]    Z←L(P COMP1 Q)R
[1]    Z←L P Q R
```

which in some APL systems is available as a primitive operator ⍥. (P COMP1 Q)¨ is equivalent to P¨Q¨, e.g.

```
      'ABC' 'XYZ'(⍋COMP1⌽)¨'CAT' 'AXE'
 2 3 1  2 1 3
```

Explicitly defining the operator draws attention to the alternative composition operator in which the left argument applies to the rightmost function:

```
[0]    Z←L(P COMP2 Q)R
[1]    Z←P L Q R
```

examples of which are:

```
      'ABC' 'XYZ'(⌽COMP2⍋)¨'CAT' 'AXE'
 3 1 2  3 1 2

      1 2(⍋COMP2⌽)¨(2 3 4)(5 6 7)
 3 1 2  2 3 1
```

5.3.2 Ambi-valency

All functions in APL2 are potentially ambi-valent and this is true also of derived functions. Consequently when writing an operator whose derived functions may be monadic or dyadic it is normal to write it in two parts, one to deal with the monadic case and the other with the dyadic case.

In Section 5.1.2 the operator SEE was defined to obtain traces for monadic processes. This can be made dyadic by

```
[0]    Z←L(P TRACE)R
[1]    →L1 IF 0≠⎕NC'L'       ⍝ branch if dyadic
[2]    →0 ⎕←'f' R '=' Z←P R   ⍝ monadic
[3]    L1:L 'f' R '=' Z←L P R  ⍝ dyadic
```

On IBM systems an alternative to the test in line 1 involves event handling thus:

```
[1]    '→L1'⎕EA'L'
```

Examples:

```
      -TRACE/ι4
3 f 4 = ¯1
2 f ¯1 = 3
1 f 3 = ¯2
¯2
```

The order of the operators reduce and **TRACE** is important:

```
      -/TRACE ι4
 f  1 2 3 4  = ¯2
¯2
```

The trace which was given step by step earlier can now be achieved by

```
      (-COM TRACE Along 1)ι4
 1  f  2  =  1
 1  f  3  =  2
 2  f  4  =  2
 2
```

5.4 Recursion with Functions and Operators

A recursive operation is one which is defined in terms of itself. A nested array is an array whose items are themselves arrays and hence is an inherently recursive structure. Using the advanced APL2 features is thus likely to bring about a shift in programming style towards recursive methods. In first-generation APL reduction of a function F applied to a vector can be described as a process whereby F is slotted in between each of the items of the vector thus:

```
    V[1] F V[2] F V[3] F ...
```

following which right to left execution takes place in the usual way. This is the iterative approach to the situation. Another equally valid way of describing reduction is to define it as

```
    (↑V) F (F/1↓V)
```

which has the merit of requiring only a description of how the first item behaves in relation to the rest, together with a (usually obvious) stopping rule to deal with the simplest case. The intermediate working is thus completely delegated to a computer. For example +/ can be described as

```
[0]    Z←SUM R
[1]    →L1 IF 1=ρ,R        ⍝ , ensures function works for scalar R
[2]    →0 Z←(↑R)+SUM 1↓R
[3]    L1:Z←R
```

Now consider the problem alluded to in Section 1.4.2 of defining a function for the path to the first occurrence of a given simple scalar L in an indefinitely deeply nested vector of vectors R. This is a recursive problem calling for a recursive solution:

```
[0]     Z←L PATH R;T
[1]     →L1 IF 1≤≡R           ⍝ branch to L1 if R not a simple scalar
[2]     →0 Z←⍳0               ⍝ if it is, stop and return empty vector
[3]     L1:T←(L∊¨∊¨R)⍳1        ⍝ identify subtree T at current depth
[4]     Z←T,L PATH T⊃R         ⍝ ..then find path within T

        V12←12(13(14 15))(16 17)

        14 PATH V12
2 2 1
```

If L does not belong to R, an error is reported. One way to deal with this is to have PATH return an empty vector in this case - this is achieved by adding another condition to line 1:

```
[1]     →L1 IF ∧/(L∊∊R),1≤≡R
```

A drawback to this solution is that it does not deal with the "level-breaker" case described at the end of Section 1.3.1.

```
        V←'ABC'(⊂⍳3)
        3 PATH V
RANK ERROR
PATH[3]    L1:T←(L∊¨∊¨R)⍳1
              ∧        ∧
```

This situation is detected when R is a non-simple scalar so an additional test must be added:

```
[0]     Z←L Path R;T
[1]     →L1 IF∧/(L∊∊R),1≤≡R
[2]     →0 Z←⍳0
[3]     L1:→L2 IF(⍳0)≡⍴R       ⍝ go to L2 if R scalar
[4]     T←(L∊¨∊¨R)⍳1
[5]     →0 Z←T,L Path T⊃R
[6]     L2:Z←(⊂⍳0),L Path↑R    ⍝ .. and return the level-breaker

        DISPLAY 2 Path V
```

The following recursive shell is one which will be used frequently in the remainder of this text:

```
        ∇Z←L FN R
[1]     →L1 IF ...            ⍝ stopping condition
[2]     →0 Z←..FN..           ⍝ recursive expression
[3]     L1:Z←...              ⍝ stopping action
```

An example of an operator developed using this shell is SIMPLE in which the function P is applied recursively to each item of R until simple arguments are reached. This has the effect of making non-pervasive functions "penetrate" deep objects.

```
[0]    Z+L(P SIMPLE)R
[1]    →L1 IF 2>≡R
[2]    →0 Z+L(P SIMPLE)¨R
[3]    L1:Z+L P R
```

```
       V12+12(13(14 15))(16 17)
```

```
  12 12    13 13   14 15    16 17
```

Read a line such as the above as an extension of 2ρR, i.e. the primary subdivision of the expression is

```
       2  (ρSIMPLE)   V12
```

with ρSIMPLE being thought of as "an enhanced version of ρ." This, like COM in Section 5.1.2, demonstrates that the binding of operators to functions is stronger than that of functions to arguments, or in simple terms operators are resolved before functions.

The recursive shell given above indicates the general organization of a recursive operation within which certain requirements must be met for it to be a *valid* recursive operation.

First, the definition must be explicit for some value or condition of the argument. This condition is the stopping condition, e.g. 2>≡R in line 1 of SIMPLE. If there is not at least one value or condition for which the definition is explicit the recursive operation is *circular* and will never terminate.

Secondly, the recursive operation must call itself with a modified argument which approaches a stopping value or condition and which it reaches in a finite number of steps. In the operator SIMPLE the recursive expression in line 2 achieves this through the **each** operator which causes the argument of the derived function to be applied to data one level down in the structure of R. A recursive operation which does not modify its argument in the course of a recursive call is called *regressive* and provided at least one recursive call is made it is also non-terminating.

In recursive operations the distinction between the actions of local and global variables is very important. Consider the following regressive recursive function:

```
[0]    Z+FN R
[1]    Z+1
[2]    →0 IF 0≥I+I-1
[3]    Z+Z,FN R
```

which terminates only when a system limit such as **WS FULL** is encountered or an attention interrupt is issued. Assuming that a value for the global variable I was set before the first call, the value of I on termination indicates how many recursive calls took place. However if the value of I were set within the function e.g. by

```
[1.1]  I+100
```

then the value of I would be reset on every call. If a temporary variable is used for an intermediate calculation at a specific level within a recursive function it *must* be localized as T is in PATH, otherwise only one copy of T would exist no matter how great the depth of the recursive calls.

Illustration : Selective Enlist

A selective enlist function performs the enlist process in a gradual way and stops when depth gets to a prescribed level L. It provides another example of a function which uses the recursive shell.

```
[0]    Z←L ENLIST R          ⍝ L is an non-negative integer, R an array
[1]    →L1 IF L≥≡R           ⍝ stop at target depth
[2]    →0 Z←↑,/L ENLIST¨R    ⍝ if not go one level lower
[3]    L1:Z←⊂R               ⍝ ensure that simple non-scalars are enclosed
```

The principle is that for vectors 0 ENLIST R is simple, and provided that the items of R are themselves vectors at every level 0 ENLIST R is equivalent to ∈R. Increasing values of L give progressively "gentler" enlists in the sense that more levels of structure are preserved.

The function ENLIST involves some sophisticated coding in line [2]. The strategy is that if the depth of the object is greater than the target, each item is separately ENLISTed and the results catenated; hence the ,/. **Catenate** reduction, ,/, requires a final enclosure in order to ensure rank reduction (this is discussed in more detail in Section 5.5.2), hence the necessity for ↑ prior to ,/. If L is greater than or equal to ≡R, ENLIST adds one additional level of enclosure.

Here is an example of the use of selective enlist applied to a tree of character strings representing the names of disk files and directories:

```
      V51←'DIR1'('F1' 'DIR2'('F2' 'F3') 'DIR3'(⊂'F4'))
      DISPLAY V51
```

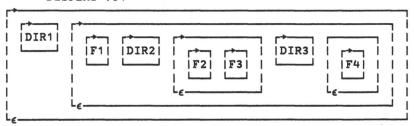

```
      ∈V51
DIR1F1DIR2F2F3DIR3F4
```

DISPLAY 1 ENLIST V51

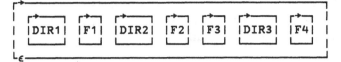

DISPLAY 2 ENLIST V51

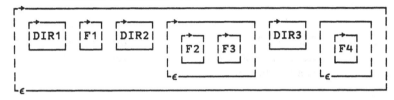

Exercises 5b

1. Construct recursive functions (a) **PRODUCT** and (b) **JOIN** which describe ×/ and ,/ in a manner akin to **SUM** in Section 5.4.

2. The function **Path** in Section 5.4 requires that the items of its right argument be vectors. What changes must be made so that the items of R may be of any rank?

3. Use the recursive shell described in Section 5.4 to write a function **CHALL** which replaces *all* occurrences of L[1] in a vector R with L[2]. (Use the function **CHANGE** of Section 1.4.2.)

4. A dyadic function P with header Z←L P R can be thought of as combining with one of its arguments, say R, to provide a new *monadic* function (P R) which is applied to the other argument, e.g. *2 can be thought of as the monadic function "square." Equally (L P) can be thought of as a monadic function applied to R so that 3* means "raise 3 to the power." Applying such functions repeatedly, say Q times, is conveniently handled by defining operators with operands P and Q.

a. Write an operator **POWER1** whose header is

[0]   Z←L(P POWER1 Q)R

and which causes P R to be applied Q times to L with the intermediate result being fed back each time. For example if P is * and R is 2, 1.5(*POWER1 3)2 means $((1.5^2)^2)^2$.

b. Write an operator **POWER2** with a similar header which causes the function
L P to be applied **Q** times to **R** with feedback so that **1.5(*POWER2 3)2** means
1.5 to the power(1.5 to the power 1.5^2). Assuming convergence, **POWER2** provides an iterative solution of the algebraic equation y =L P y.

c. Use **POWER2** to find a solution of the equation y = cos(y). Take a start
value of 1 and investigate the number of iterations required for convergence to
six significant figures.

d. A cryptographer defines the 26 upper-case letters of the alphabet as **ALF**
and uses an anagram of **ALF**, e.g. **ALF[26?26]**, as a code replacement string **L**
to encrypt a message **R**. The function he uses to do this is

```
[0]     Z+L CODIFY R
[1]     Z+L[ALFιR]
```

To make his encryption more secure he repeatedly encrypts the encrypted word.
Which of **POWER1** and **POWER2** above should he use in order to encrypt a source
message four times in succession? How does the receiver then decode it?

5. To "polish" a matrix means to subtract a sequence of values, say the row
means, one from every row, and then another sequence, say the column means,
one from every column. Write an operator **POLISH** which achieves this, and use
it to obtain the mean polish and also the median polish of the matrix

```
    0   6   6
    4   0   2
```

(The median of a vector **R** is defined as **.5×+/R[⌈.5×0 1+ρR+R[⍋R]]**)

5.5 Extensions to First-generation APL Operators

5.5.1 Reduction

A general principle of reduction is that it reduces rank. It does not reduce depth.
With scalar dyadic functions this is an entirely natural rule, e.g. +/2 2 2 transforms a vector to a scalar and and +/2 2ρι4 transforms a matrix to a vector.
However when reduction is applied to non-pervasive functions, adjustments to
depth must sometimes be made in order to maintain the rank-reduction rule.
The functions ρ and , have the property of increasing rank, e.g. starting with
two scalars 2 and 4 both 2ρ4 and 2,4 give results which are vectors. In order
that ρ/2 2 2 and ,/2 2 2 should produce scalars an extra level of nesting
must be provided.

As noted in Section 5.4 F-reduction can be described by inserting the function F into all the available spaces of V[1]   V[2]   ... and evaluating the
resulting expression. So what changes must be made to the first-generation
APL rule to deal with this state of affairs? Since indexing provides *cross-sections*
of arrays, V[1] is not the first item of V, rather it is a *container* for the first item
which can be opened by ⊃. Thus it is F¨ rather than F which is inserted into the
spaces of V[1]   V[2]   .... At the function phase the **each** rule (see Section
4.1) applies. If the items are scalars or if F is pervasive the **each** makes no difference, and so there is no inconsistency with the first-generation APL view of
reduction.

Alternatively one can think in terms of **pick** which penetrates the items, so
that F-reduction is obtained by inserting F into the spaces of

      (1⊃V)   (2⊃V)   (3⊃V) ....

and applying a final enclosure.

Illustration : Reduction applied to matrix multiplication

Consider the sequence of algebraic matrix multiplications which is given by

      +.×/A B C

where A, B and C are compatible matrices. One way to determine the exact
result of this expression is to consider a recursive definition of the derived function DF arising from applying reduction to +.× (cf. SUM in Section 5.4):

```
[0]     Z←DF R
[1]     →L1 IF 1=ρR
[2]     →0 Z←c(↑R)+.×⊃DF 1↓R
[3]     L1:Z←c↑R
```

A B C is a nested vector, comprising cells which *contain* matrices. The function
+.× can properly be applied only to items, that is the *contents* of cells, hence the

depth reducing ↑ and ⊃ in line 2. After executing the +.× the resulting matrix occupies a single cell, hence the enclosure in line 2. The **encloses** which appear immediately to the right of the assignment statement in both lines 2 and 3 show that the result of the +.× reduction is a scalar of depth two. In order to achieve the mathematical matrix product **ABC** which is a depth-one matrix it is necessary to apply either ⊃ or ↑. The expressions

⊃+.×/A B C and ↑+.×/A B C

have equivalent output for the reasons given in the Section 4.3.4. Another way of looking at the role of ⊂ and ⊃ in DF is that what reduction reduces is *rank*. ρA B C is 3, and has rank one since ρ always returns a vector, and so the result of the +.× reduction must be a scalar, namely the *enclose* of the solution matrix. Eliding the references to R in the recursive part (line 2) gives

Z←⊂..⊃DF

which is another occurrence of the ⊂ ⊃ sequence observed in the **each** rule.

5.5.2 Reduction with Rank Greater than one

If reduction is applied to objects of rank two, enclosure takes place along the last dimension and the vector rule for reduction is applied to each item of the result. Enclosure along the last axis gives a row vector whose items are the rows of the array. The final result is the vector whose items are the **plus** reductions of each of them, e.g.

    DISPLAY +/2 3ρι6

Now consider the reductions of non-pervasive functions such as ρ and ,.

    DISPLAY ,/2 3ρι6

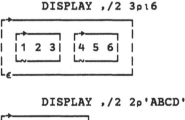

    DISPLAY ,/2 2ρ'ABCD'

          DISPLAY ⊃¨,/¨⊂[2]2 3ρ⍳6

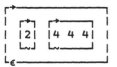

,/A is the same as ⊂[2]A for simple arrays A. Contrast this with

          DISPLAY ,/¨2 2ρ'ABCD'

The **each** makes a scalar function from ,/, and so following the discussion in
Section 4.1.1 ,/ is applied to the four character scalars separately, and the
result is a simple two by two matrix.
   In the next example enclosure again gives two row vectors:

          DISPLAY ρ/2 2ρ⍳4

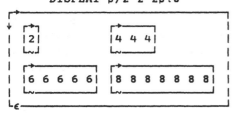

and the result is (1ρ2)(3ρ4). The principle, formally defined as

          F/A ↔ ⊃¨F/¨⊂[ρρA]

extends in a natural way to arrays of higher dimension:

          DISPLAY ρ/2 2 2ρ⍳8

and the result is (1ρ2)(3ρ4).

In the above example the principles of rank reduction apply and the result is a
two by two matrix. Enclosure along the last dimension gives a two by two structure of row vectors, and applying ρ/ to each gives

          (1ρ2) (3ρ4)
          (5ρ6) (7ρ8)

A similar argument applies with **catenate** in the next example:

DISPLAY ,/2 2 2ρ'ABCDEFGH'

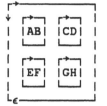

the steps of which can be broken down as:

DISPLAY ⊃¨,/¨⊂[3]2 2 2ρ'ABCDEFGH'

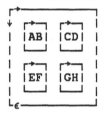

5.5.3 Scan

Scan defines an *array* of reductions, and informally therefore preserves the rank which reduction reduces. For example

    V←+.×\M1 M2 M3

for compatible matrices M1, M2 and M3 is a depth two vector of matrices

    (M1) (↑+.×/M1 M2) (↑+.×/M1 M2 M3)

i.e. V[1]≡⊂M1, V[2]≡⊂↑+.×/M1 M2, V[3]≡⊂↑+.×/M1 M2 M3. Because V is a vector, ↑V and ⊃V are not equivalent in this case. Both bring about depth reduction but ↑ returns the matrix M1, whereas ⊃ returns a rank three depth one array whose planes are M1, (M1+.×M2), and (M1+.×M2+.×M3) respectively, possibly padded with 0s.

Illustration : Co-ordinates of Spirals

The initial point of a spiral drawn as a two dimensional graph using Cartesian co-ordinates and O as origin is taken to be P(0,1). A function SPIRAL defines four new points which are generated by rotating OP through an angle of R anti-clockwise degrees, and stretching it by a factor L. The result of SPIRAL is a matrix, each row of which is the co-ordinates of a point on the spiral. The aux-iliary function ΔSPIRAL generates in its second line the rotation matrix M:

$$\cos \theta \quad \sin \theta$$
$$-\sin \theta \quad \cos \theta$$

for which **V+.×M** gives the co-ordinates of the result of rotating the point with co-ordinates **V** through an angle θ

```
[0]    Z+L SPIRAL R           ⍝ L is stretch factor, R an angle in degrees
[1]    Z←⊃+.×\(⊂0 1),4⍴⊂L ∆SPIRALOR÷180

[0]    Z+L ∆SPIRAL R;S;C
[1]    (S C)←1 2OR            ⍝ S and C are sin and cos
[2]    Z←L×2 2⍴C S(-S)C       ⍝ Z is the transformation matrix
                                 to move to the next point
```

```
        2 SPIRAL 45
 0          1
⁻1.414     1.414
⁻4         0
⁻5.657    ⁻5.657
 0        ⁻16
```

Illustration : Scans with Binary Arguments

With the exception of **circle** the scans of the scalar dyadic functions have some interesting properties. A useful binary matrix for demonstrating these is constructed by

```
        M53←(8⍴2)⊤15,(6⍴51 43),113

        M53
0 0 0 0 0 0 0 0
0 0 0 0 0 0 0 1
0 1 1 1 1 1 1 1
0 1 0 1 0 1 0 1
1 0 1 0 1 0 1 0
1 0 0 0 0 0 0 0
1 1 1 1 1 1 1 0
1 1 1 1 1 1 1 1
```

A better visual way of representing this matrix is to represent the **0**s with dots and the **1**s with asterisks:

```
        '.*'[1+M53]
........
.......*
.*******
.*.*.*.*
*.*.*.*.
*.......
*******.
********
```

Here are the scans of the six relational and four logical primitive functions applied to this matrix:

```
      =\              <\              ≤\              ≥\              >\
  .*.*.*.*        ........        .*******        .*.*.*.*        ........
  .*.*.*..        .......*        .*******        .*.*.*..        ........
  ........        .*......        .*******        ........        ........
  ..**..**        .*......        .*******        ........        ........
  *..**..*        *.......        *.******        ********        ********
  *.*.*.*.        *.......        *.******        ********        ********
  *******.        *.......        *******.        ********        *.*.*.**
  ********        *.......        ********        ********        *.*.*.*.

      ≠\              ∨\              ∧\              ⍲\              ⍱\
  ........        ........        ........        .*.*.*.*        .*******
  .......*        .......*        ........        .*.*.*..        .*******
  .*.*.*.*        .*******        ........        ..******        .*******
  .**..**.        .*******        ........        ..******        .*******
  **..**..        ********        *.......        *.......        **......
  *******.        ********        *.......        *.......        **......
  *.*.*.**        ********        *******.        *.......        *.*.*.**
  *.*.*.*.        ********        ********        *.......        *.*.*.*.
```

There are four scans which between them have the greatest practical use when applied to binary vectors. Subject to the universal rule that scan leaves the first item unchanged, the behavior of these scans can be summarized:

  ∨\ : detects the first 1 and switches all following bits to 1
  ∧\ : detects the first 0 and switches all following bits to 0
  <\ : detects the first 1 and switches all following bits to 0
  ≤\ : detects the first 0 and switches all following bits to 1

Illustration : Delete leading blanks from a character vector

This can be achieved by either of two expressions, viz:

```
(∨\' '=CV)/CV
```

or

```
(~∧\' '≠CV)/CV
```

Illustration : Display comments only on an APL line

This can also be achieved by either of two expressions, viz:

```
(∨\'⍝'=LINE)/LINE
```

or

```
(~∨\'⍝'≠LINE)/LINE
```

Illustration : Remove first occurrence only

(a) character from a character vector.

To remove the first occurrence only of X from a character vector use either of the two expressions:

    `(≤\'X'≠CV)/CV`

or

    `(~<\'X'=CV)/CV`

(b) word from a sentence.

Define the sentence as a vector of character vectors (words):

    `V52←'VERMONT' 'IN' 'THE' 'THE' 'FALL'`

The first occurrence of THE can be removed by either of

    `(≤\~V52=̈⊂'THE')/V52`

or

    `(~<\V52=̈⊂'THE')/V52`

The above illustrations exhibit a duality inherent in the scans listed above. In particular the last two show that the dual of < is ≤ and not > as intuition might suggest. Another way of describing the behavior of the four scans in a concise way is by the following table:

	0-continuation	1-continuation
0-detector	∧	≤
1-detector	<	∨

The functions ∧ and ∨ down the leading diagonal have the property of idempotency, that is

    A ↔ A∧A   and   A ↔ A∨A

Consider the functions which are the "not"s of the functions in the above table. The behavior of their scans depends on whether the first bit is 1 or 0, and their effect is either that of an "alternator," that is a function which takes a series of uniform bits and transforms it into an alternating sequence of 1s and 0s, or a "sweeper," that is a function which makes all bits alike.

For first bit = 1 the following table applies:

	0-continuation	1-continuation
Alternators	>	⩘
Sweepers	⩘	≥

If the first bit is 0 the roles of alternator/sweeper are reversed. The functions down the non-leading diagonal of this table are cyclic of order two, that is

$$A \leftrightarrow A \geq A \geq A \quad \text{and} \quad A \leftrightarrow A > A > A$$

The functions = and ≠ are also alternators but are not dependent on the first bit. Instead they have the effect of doubling the length of the subsequences within alternating sequences, and hence quadrupling, octupling etc. them by repeated application.

Illustrations : Spacing character vectors

Spaces can be placed between alternate characters of a character vector by:

```
      (=\(2×⍴T)⍴0)\T←'FREDERICK'    ⍝ start with space
F R E D E R I C K
      (≠\(2×⍴T)⍴1)\T←'FREDERICK'    ⍝ start with first character
F R E D E R I C K
```

Selecting alternate items

Using scan is an alternative method to indexing:

```
      (=\(⍴T)⍴0)/T←⍳10              ⍝ even numbered items
2 4 6 8 10
      (≠\(⍴T)⍴1)/T←⍳10              ⍝ odd numbered items
1 3 5 7 9
```

Illustration : Adding columns of zeros to table

```
      (≠\(2×2⊃⍴NM)⍴1)\NM
```

opens up alternately spaced columns of zeros in a numeric matrix NM.

Illustration : Parity checking

=/BV and ≠/BV give 1-parity and 0-parity checks respectively for a binary vector BV:

```
      BV←0 1 1 1 0 1
      (=/BV),(≠/BV)
1 0
```

Further ≠\BV gives on-going 0-parity checks on the sequence so far:

```
      BV←0 1 1 1 0 1
      ≠\BV
0 1 0 1 1 0
```

The scan diagrams given earlier cover the six relational and four logical scalar dyadic functions. Apart from the circle function (o) which is special, there are ten further primitive scalar dyadic functions. What are the results of applying their scans to the matrix M53? Four of them give results outside the binary domain, the remaining six duplicate the tables in the following pairings:

$$(< \mid) \quad (≤ !) \quad (≥ *) \quad (∨ ⌈) \quad (∧ ⌊) \quad (∧ ×)$$

The functions above can also be arranged in dual pairs where "dual" in this context means that ~F\~V is equivalent to (dual F)\V. The primitive functions which possess duals are

```
F      =  ∧  ⍲  <  >  ⌈  ⌈  |
dual   ≠  ∨  ⍱  ≤  ≥  ⌊  ×  !
```

The easiest way to visualize this duality is to rotate the appropriate scan matrices above about a horizontal axis.

5.5.3.1 Reversing scans

The following operators invert scans with vector arguments:

```
[0]    Z←(P UNSCAN)R
[1]    Z←R[1],(1↓R)P ¯1↓R

[0]    Z←(P UNDO)R
[1]    Z←R[1],(¯1↓R)P 1↓R
```

and the following relations apply:

```
P COM UNDO R      ←→     P UNSCAN R
P COM UNSCAN R    ←→     P UNDO R

[0]    Z←L(P COM)R
[1]    Z←R P L
```

UNSCAN works if P is associative and there exists an inverse function (see Section 4.4.1.2) - the only functions satisfying this criterion are + - = and ≠.

```
    -UNSCAN   and -COM UNDO both reverse  +\
    ÷UNSCAN   and ÷COM UNDO both reverse  ×\
    =UNSCAN   and =UNDO both reverse  =\
    ≠UNSCAN   and ≠UNDO both reverse  ≠\
```

Illustration : Gray codes

Gray codes are a method of representing integers using binary digits in such a way that only one bit is changed when an integer is incremented by 1. ≠\ converts Gray code representations to binary and ≠\UNSCAN does the reverse. Gray codes can therefore be obtained by first obtaining binary representations using ⊤ and then applying UNSCAN. A table of the first 15 integers in Gray code is given by:

```
      T,⊃[2]≠UNSCAN¨⊂[1](4ρ2)⊤T←ι15
 1  0  0  0  1
 2  0  0  1  1
 3  0  0  1  0
 4  0  1  1  0
 5  0  1  1  1
 6  0  1  0  1
 7  0  1  0  0
 8  1  1  0  0
 9  1  1  0  1
10  1  1  1  1
11  1  1  1  0
12  1  0  1  0
13  1  0  1  1
14  1  0  0  1
15  1  0  0  0
```

5.5.4 Expand

Suppose that a character matrix M is given:

```
      M←2 3ρ'BATMAN'
      DISPLAY M
┌→──┐
↓BAT│
│MAN│
└───┘
```

together with the instruction "Space the matrix M." On the structural level this might imply using **ravel with axis** (see Section 1.2.3), e.g.

```
      M←2 3ρ'BATMAN'
      DISPLAY ,[1.1]M
┌┌→──┐
↓↓BAT│
││   │
││MAN│
└└───┘
```

At the data level there are at least nine possible interpretations of this instruction as the following set of expressions show. Consider first

```
DISPLAY 1 0 1 0 1\M
```

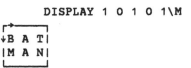

Unlike first-generation APL where \ is the *function*, **expand**, \ is now an *operator* so the derived function in the above expression is 1 0 1 0 1\. Enclosing M forces scalar expansion of the right argument as in the next expression:

```
DISPLAY 1 0 1 0 1\⊂M
```

```
DISPLAY 1 0 1 0 1\¨M
```

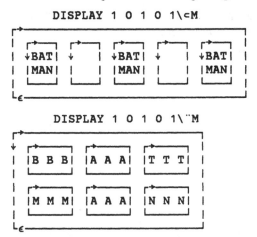

Replacing **enclose** by **each** forces scalar expansion of *each item* of M:

The **each** in the above expression has the derived function 1 0 1 0 1\ as its operand. As with **replicate** (see Section 2.2.1) the left operand of \ must be simple. An expression such as (⊂1 0 1 0 1)\¨M thus leads to a **DOMAIN ERROR**. To apply **each** in this situation it is necessary to create a defined function, e.g.

```
[0]    Z←L EXPAND R
[1]    Z←L\R
```

```
       DISPLAY (1 0 1)(1 0 0 1)\¨'AB' 'DE'
DOMAIN ERROR
       DISPLAY(1 0 1)(1 0 0 1)\¨'AB' 'DE'
       ∧                      ∧
       DISPLAY (1 0 1)(1 0 0 1)EXPAND¨'AB' 'DE'
```

```
 ┌→──────────────┐
 │ ┌→──┐ ┌→────┐ │
 │ │A B│ │D  E │ │
 │ └───┘ └─────┘ │
 └∊──────────────┘
```

DISPLAY (⊂1 0 1 0 1)EXPAND¨M

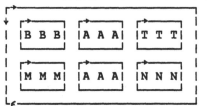

Partial enclosure requires matching lengths between 1s in the left operand and the number of columns or rows in the right argument:

DISPLAY 1 0 1 0 1\⊂[1]M

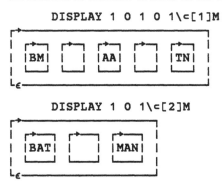

DISPLAY 1 0 1\⊂[2]M

Now use **each** once again to force itemwise scalar expansion of first columns, then rows:

DISPLAY 1 0 1\¨⊂[1]M

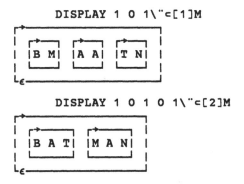

DISPLAY 1 0 1 0 1\¨⊂[2]M

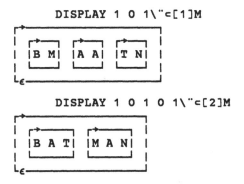

Matching disclosures make the results simple:

```
        DISPLAY ⊃[2]1 0 1\"⊂[1]M
┌→──┐
↓B M│
│A A│
│T N│
└───┘
        DISPLAY ⊃[1]1 0 1 0 1\"⊂[2]M
┌→─┐
↓BM│
│  │
│AA│
│  │
│TN│
└──┘
```

5.5.5 Outer Product

The key phrase associated with outer product is, as in first generation APL, *each with every*, that is *each* of the items in the left argument is combined through the function operand with *every* item in the right argument. However since any result producing dyadic function can be the operand, the depth of the result may change. Consider for example

```
        DISPLAY (2 2)3∘.+3(4 5)
```
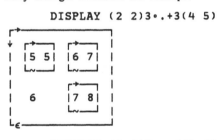

```
        DISPLAY (2 2)3∘.ρ3(4 5)
```
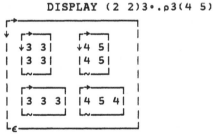

Each cell of the result is obtained as

$$Z[I;J] \leftrightarrow ⊂(⊃L[I])F⊃R[J]$$

Compare this with the **each** rule (see Section 4.1) which states that if $Z \leftarrow L \ F"R$ then

$$Z[I] \leftrightarrow ⊂(⊃L[I])F⊃R[I]$$

If the arguments are of higher rank than vectors, replace replace I and J above by the indices necessary to reach scalar level. For example, if L and R are matrices

```
Z[I;J;K;L] ←→ ⊂(⊃L[I;J])F⊃R[K;L]
```

or more generally

```
Z[LI;RI] ←→ ⊂(⊃L[LI])F⊃R[RI]
```

where LI and RI are index sets for L and R of appropriate rank.

The shape rule from first-generation APL still applies, viz.:

the shape of the result is the catenation of the shapes of the arguments,

or more formally:

```
ρZ ←→ (ρL),ρR
```

Establishing the shape of the result initially is often very helpful in working out the values of outer products. For example, the following outer products necessarily result in two-item vectors:

```
DISPLAY¨(1 2∘.×⊂3 4)((⊂1 2)∘.×3 4)
```

On the other hand, the following outer product must be a scalar :

```
DISPLAY (⊂1 2)∘.×⊂3 4
```

5.5.6 Inner Product

In first-generation APL, inner product operands are restricted to primitive scalar functions and the shape vector rule dominates the outcome in that if L and R are left and right arguments respectively it is necessary (subject to scalar-extension flexibility) that

```
(⁻1↑ρL) ←→ 1↑ρR
```

The shape of the result is (ρL),ρR with both the matching inner shape vector items removed.

The most common form of inner product is that in which a pair of matrices L of dimension (m,k) and R of dimension (k,n) is reduced to a single matrix of dimension (m,n).

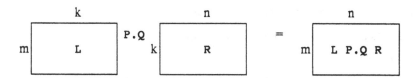

Each cell of the result of L P.Q R is the result of first applying the functions Q between **each** item of a pair of vectors, one a row of L and the other a column of R, and then doing a P reduction on the result. The two functions P and Q thus behave quite differently, Q is a function operating between matching pairs of items, P is the operand of a reduction.

The most frequently occurring inner product is +.× which is equivalent to matrix multiplication in the mathematical sense. Each cell of the result is an itemwise product of two vectors, and then **plus** reduction is applied to the resulting vector.

Another frequent inner product in first-generation APL is ∧.=. By the same reasoning this gives 1 only if **all** (∧/) the matching pairs in the vectors are equal. Similarly ∨.= gives 1 if **at least one** (∨/) of the matching pairs are equal. ∧ and ∨ as left operands of inner products thus model the universal and existential quantifiers respectively of symbolic logic.

The logical functions give rise to other inner products with binary arguments:

∧.∨   gives 1 if all pairs contain at least one 1
∨.∧   gives 1 if at least one pair has two 1s
∧.⍲   gives 1 if there are no pairs of matching 1s
∧.⍱   gives 1 if all the pairs consist of two 0s

Some inner products which apply to numeric arguments are:

⌈.⌊   gives the maximum of a set of pairwise minima (maximin)
⌊.⌈   gives the minimum of a set of pairwise maxima (minimax)
⌊.-   gives the minimum of a set of differences of paired items

In APL2 the shape rule still applies but operands may be both user-defined functions on the one hand, and non-scalar primitive functions on the other. For example in considering the last of the above inner products it is likely that the **absolute** difference might be of more interest, that is the inner product ⌊.AD where

```
[0]     Z+L AD R
[1]     Z+|L-R

        2 4 7 9 ⌊.AD 4 1 6 15
1
```

The price of this increase in flexibility is a slight increase in the complexity of the inner product rules. To evaluate L P.Q R the following sequence of actions must be carried out:

Step 1 : Enclose L and R along inner matching axes.
Step 2 : Perform Q outer product.

Step 3 : Apply `P/` within each cell, or equivalently `P/¨` to each cell.

Consider as a further example the inner product

```
T+.pT+2 2p14
```

Enclosure along last and first dimensions of left and right arguments respectively means that the `p` step of the operation consists of forming the outer product of the two vectors

```
(1 2)(3 4)   and   (1 3)(2 4)
```

The **each** rule applied to outer products as described in the previous section leads to the depth-two rank-two array whose four cells are

(⊂1 2ρ1 3)	(⊂1 2ρ2 4)
(⊂3 4ρ1 3)	(⊂3 4ρ2 4)

Now apply `+/¨` (that is `+/` *within* each cell) to give the final result

```
4       6
8 8 8   12 12 12
```

Eliding arguments the picture is

emphasizing that a function composition occurs *within* each cell.

Formally the definition of the inner product `Z+L(P.Q)R` is

```
Z+F/¨(⊂[ppL])•.G ⊂[1]R
```

and the shape of its result is

```
pZ ↔ (¯1↓pL),1↓pR
```

Inner products allow great programming versatility as the next illustrations show.

Illustration : Finding vowels in words

Consider the difference between the following two expressions:

```
(⊂'CAT')•.ι'AEIOU'
2 4 4 4 4
'CAT'⌊.ι'AEIOU'
2
```

The inner product `⌊.ι` returns the index of the first vowel in the word `'CAT'`. To find the first vowel in each of a vector of words use

```
      'CAT'  'ELK'⌊.ι¨⊂'AEIOU'
2  1
```

or equivalently

```
      ⌊/'CAT'  'ELK'∘.ι'AEIOU'
2  1
```

The case where the right argument of `P.Q` is a scalar is of special interest since `P/` of a scalar does not involve an execution of `P`. Thus if `P` is *any* scalar function

```
      'CAT' P.ι'A'
```

is equal to 2 and

```
      'CAT' P.ι⊂'AEIOU'
```

is equal to 4.

Illustration : Gradient of mid-points

Define

`[0]`	`Z←L GRAD R`	`[0]`	`Z←L MIDPT R`
`[1]`	`Z←⌹L⍪.-R`	`[1]`	`Z←.5×L+R`

to return the gradients and mid-point of pairs of points defined as two-item vectors of Euclidean co-ordinates. Ignoring the complexities of zero and infinite gradients, if **A**, **B** and **C** are three points then

```
      A B GRAD.MIDPT B C
```

gives the gradient of the line joining the midpoints of AB and BC.

Illustration : Sampling Extreme Values from Uniform Distribution

This illustration is a variation on the function **deal**. The expression n?100 describes a random sample of n integers drawn from the uniform distribution of integers 1 to 100. For n not exceeding 100, n⌈.?100 returns the maximum of a sample of n such integers. n⌈.?¨m⍴100 returns the maxima and n⌊.?¨m⍴100 the minima of m such samples. For example:

```
      10⌈.?¨15⍴100
97 95 88 92 90 97 93 95 100 84 56 97 95 96 94
      3⌈.?¨15⍴100
84 57 79 81 81 67 95 93 95 53 72 85 69 58 92
```

```
      10L.?¨15ρ100
5 2 10 7 9 16 11 4 1 2 1 1 3 3 7
      3L.?¨15ρ100
19 56 17 24 12 44 22 20 45 18 44 28 34 16 14
```

The functions , and ρ lead to a further subtlety on account of their rank-increasing property discussed in Section 5.5.1. Consider for example the inner product 1 2 3,.,4 5 6. The shape rule for inner product requires that the result is a scalar since discarding the inner (and only) axes leaves nothing in the shape vector. To find its depth and value determine first the **catenate** outer product of the two scalars obtained by enclosure along the inner (and only) axes:

DISPLAY (⊂1 2 3)∘.,⊂4 5 6

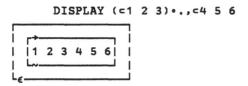

Then the ,/¨ corresponding to the leftmost **catenate** in the inner product results in a further enclosure for the reason given above, giving as the final result the depth three scalar:

DISPLAY 1 2 3,.,4 5 6

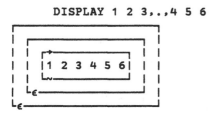

In summary the evaluation of inner product requires an application of several important identities all of which play a role in determining the data, shape and structure of the result. Formally these are:

1. L P.Q R ↔ P/¨(⊂[ρρL]L)∘.Q⊂[1]R
2. For Z←L∘.P R, each item Z[I;J] ↔ ⊂(⊃L[I])Q⊃R[J]
3. P/¨A ↔ ⊂P/⊃A .

As a further example consider the evaluation of the expression 1 2 3ρ.ρ4 5 by following the formal rules. First by identity 1

    1 2 3ρ.ρ 4 5 ↔ ρ/¨(⊂1 2 3)∘.ρ⊂4 5

There is no need for axis specification on the enclosures since both are vectors. The shape of the outer product (⊂1 2 3)∘.ρ⊂4 5 is the join of two ιOs and so the outer product itself is a scalar. Applying identity 2 to each item in the outer product - in this case the only item - gives

    ⊂(⊃1 2 3)ρ(⊃4 5)

the result of which is:

         DISPLAY ⊂(⊃1 2 3)ρ(⊃4 5)

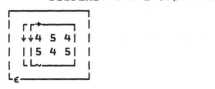

that is a scalar containing a rank 3 array. Now ρ/¨ is applied to this interim
result. Applying identity 3, ρ/¨⊂(⊃1 2 3)ρ(⊃4 5) can be replaced by
⊂ρ/⊃⊂(⊃1 2 3)ρ(⊃4 5). Simplifying the ⊃s in this expression gives

         DISPLAY ⊂ρ/1 2 3ρ4 5

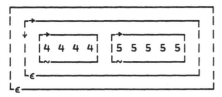

Finally then, (⊂1 2 3)ρ.ρ(⊂4 5) is equivalent to

         DISPLAY 1 2 3 ρ.ρ 4 5

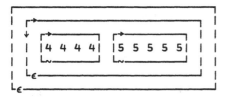

Exercises 5c

1. Given

         H←2 2ρ(2 1ρ⍳6)(2 3ρ7+⍳6)(2 4ρ⌽⍳6)(2 2ρ9)

what are ,/H and ⊂[2]H ?

2. If

    M←2 3ρ'ABCDEF'
    A←2 2 3ρ'ABCDEFGHIJKL'

   what are the values of the following

   a. ,/M   b. ,/¨M   c. ,/A   d. ,/¨A ?

3. a. Write a function SUBMAT which returns every consecutive submatrix of
shape L occurring within a matrix R. For example if M54 is the matrix

```
1 0 1 0 1
1 1 0 1 1
0 1 1 0 1
1 1 0 1 0
```

`3 3 SUBMAT M54` should return the 2x3 matrix of consecutive 3x3 submatrices occurring in `M54`:

```
1 0 1   0 1 0   1 0 1
1 1 0   1 0 1   0 1 1
0 1 1   1 1 0   1 0 1

1 1 0   1 0 1   0 1 1
0 1 1   1 1 0   1 0 1
1 1 0   1 0 1   0 1 0
```

(Hint - you may find **n-wise reduction** useful, see Section 2.2.2)

    b. Use `SUBMAT` to detect every occurrence of the pattern

```
    1
1 0 1
    1
```

in a bit matrix.

    c. Write a function `PATIN` which generalizes this process to match any given binary pattern in any binary matrix.

4. The following three exercises all involve the use of **scan**.

    a. Write an expression which returns a given character vector `CV` with double spacing between each item, that is two spaces should follow every character, e.g.

```
F   R   E   D   E   R   I   C   K
```

    b. Write an expression which returns `CV` written in blocks of two characters, each followed by a space, e.g.

```
FR ED ER IC K
```

    c. Write an expression which deletes a comment from an APL line, that is all characters to the right of `⍝` including `⍝` itself.

5. a. Predict the value and structure of

```
((2 2)3)∘.⍴6(4 1⍴'ABCD')
```

    b. For the two simple matrices

```
A←2 2⍴⍳4
B←2 2⍴⌽⍳4
```

evaluate in full detail the inner product A+.×B and confirm that the result is the same as in first-generation APL.

6. Evaluate the following outer products in terms of value, shape and structure:

a. 2 4•.+1 4 6   d. 2 3•.ρ1 4
b. 2 4•.,1 4 6   e. 2 4 6•.,'AB'
c. 2 2•.ρ1 4   f. 2 4 6•.,'AB' 'CDE'

7. Using an analogous argument to that in Section 5.5.6 for 1 2 3,.,4 5 6 determine without using a computer the shape, structure and value of:

a. 3 2 1ρ.ρ3 2 1   d. 1 2 3ρ.,4 5 6
b. 3 2 1ρ.ρ3 2 1   e. 1 2 3~.+2 3 4
c. 1 2 3,.ρ4 5 6   f. 1 2 3+.~2 3 4

8. This exercise is designed to force precise application of the rules for reduction and inner and outer products, and should therefore be done in the first place *without* help from a computer.

   The two functions AVG and MID which follow both return the average of L and R in the particular case where L and R are both simple numeric scalars.

```
[0]    Z+L AVG R              [0]    Z+L MID R
[1]    Z+.5×+/L,R             [1]    Z+L+.5×-/R,L
```

Use the rules for reduction and inner and outer products to find the values of

a. AVG/ι4   MID/ι4
b. 1 2•.AVG 3 4 5   1 2•.MID 3 4 5
c. (⊂1 2)•.AVG 3 4 5   (⊂1 2)•.MID 3 4 5
d. 1 2,.AVG 3 4 5   1 2,.MID 3 4 5
e. 1 2 AVG.,3 4 5   1 2 MID.,3 4 5
f. 1 2 AVG.MID 3 4 5   1 2 MID.AVG 3 4 5?

5.5.7 Further Topics on Inner and Outer Products

Illustration : Sequences of Inner Products

In APL2 operands may be either derived or user-defined functions, so that expressions such as +.×.- which were invalid in first-generation APL now have meaning, e.g.

      1 2+.×.-3 4   (a)
4

      1 2+.(×.-)3 4   (b)
4

To evaluate these apply the binding rules first (see Section 5.2) to work out the order in which the operators are applied and then consider the structure rules as the first step towards evaluating the detailed results.

Looking in detail at (a), the binding rules, or equivalently the rule that operators have long left scope, show that the derived function of the the leftmost inner product becomes the left operand of the rightmost inner product and so the final derived function is (+.×).-. The first-generation APL rule suggests an answer

      (+.×)/(1-3),(2-4)   ↔   (+.×)/⁻2 ⁻2   ↔   4

Under the APL2 rule the first step is to obtain the outer product

      DISPLAY (⊂1 2)∘.-⊂3 4

```
┌───────────┐
│ ┌→────┐   │
│ │⁻2 ⁻2│   │
│ └~────┘   │
└∈──────────┘
```

Then apply +.×/ within each cell to obtain

      DISPLAY +.×/¨(⊂1 2)∘.-⊂3 4

4

The APL2 rule thus follows first-generation intuition in this case, the difference being that for correct evaluation it is necessary to think of **enclose** and **each**, even although neither was present in the original expression.

In (b), following first-generation APL intuition, one might suppose that the derived function ×.- was applied first between two pairs of scalars and so should be equivalent to -, since if the arguments of an inner product P.Q are scalars then the function P plays no part, i.e. P.Q is equivalent to Q. This reasoning would lead to a final result

      1 2+.-3 4

namely ⁻4. Correct application of the APL2 rules, however, leads to an initial outer product

```
DISPLAY (⊂1 2)∘.(×.-)⊂3 4
```

4

which by the shape rule for the outer product is a scalar, and so it is the +, not the ×, which is the null function. First-generation APL intuition is thus misleading in this case.

Illustration : Inner Products with Nesting

This illustration is a discussion of the differences between a pair of expressions which might at first sight look as if they should give the same results. They are

```
(⊂1 2)+.×⊂3 4 and     +.×/(⊂1 2),⊂3 4
```

To evaluate

```
(⊂1 2)+.×⊂3 4                                          (a1)
```

start with the shape rule which requires that the final result is a scalar. Steps 1 and 2 lead to:

```
DISPLAY (⊂⊂1 2)∘.×⊂⊂3 4
```

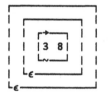

which is too deep for +/ to have an effect at Step 3. The final result is therefore

```
DISPLAY +/¨(⊂⊂1 2)∘.×⊂⊂3 4
```

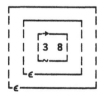

On the other hand consider

```
+.×/(⊂1 2),⊂3 4                                        (a2)
```
11

The rank rule for reduction (see Section 5.5.1) shows that this must be a scalar, namely

```
      (⊂1 2)+.×¨⊂3 4
11
```

(a1) and (a2) are thus not equivalent.

Illustration : Displacement Vectors

Let **B** defined as

```
      B←2 2ρ(0 1)(2 0)(1 0)(0 2)
      DISPLAY B
```

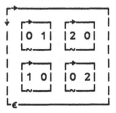

be considered as a matrix of displacement vectors in two dimensional space (or forces, velocities, etc.) so that **B** is

$$v_1 \quad v_2$$
$$w_1 \quad w_2$$

and v_1 is (0,1), v_2 is (2,0) and so on. Then `⊃1 2+.×B` gives $(v_1+2w_1)(v_2+2w_2)$ and `⊃(⊂1 2)+.×B)` gives $(v_1'+w_1')(v_2'+w_2')$ where v_1' and v_2' represent v_1 and v_2 with x- and y- "stretch factors" 1 and 2 respectively applied to each displacement.

Consider first

```
      DISPLAY 1 2+.×B
```

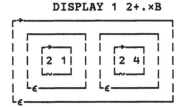

The result is clearly `(1×B[1;]) + (2×B[2;])`, but why are the vectors doubly enclosed? To find out, follow the three steps for evaluating inner products in detail:

Step 1:

DISPLAY ⊂[1]B

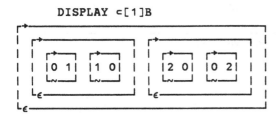

Step 2:

DISPLAY (⊂1 2)∘.×⊂[1]B

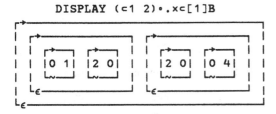

Step 3:

DISPLAY +/¨(⊂1 2)∘.×⊂[1]B

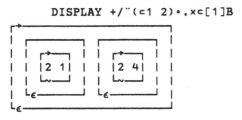

Next consider (⊂1 2)+.×B .

Steps 1 and 2: form outer product (⊂⊂1 2)∘.×⊂[1]B whose two items are

DISPLAY¨((⊂1 2)×1⊃⊂[1]B)((⊂1 2)×2⊃⊂[1]B)

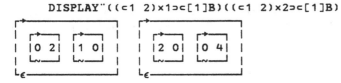

Step 3: apply +/ to each of these 2 cells, or equivalently +/¨ to the entire outer product to give the final result

DISPLAY (⊂1 2)+.×B

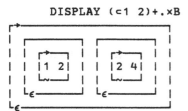

This result can also be written (⊂(⊂1 2)×B[1;]) + (⊂(⊂1 2)×B[2;]).

Illustration : Outer and Inner Products with Explicit Each

What are:

```
(⊂1 2) ×¨ ⊂3 4                    (a)

(⊂1 2) ∘.×¨ ⊂3 4                  (b)

(⊂1 2) +.(×¨) ⊂3 4                (c)

(⊂1 2) +.×¨ B                     (d)
```

where B←2 2ρ(0 1)(2 0)(1 0)(0 2) as in the previous illustration?

In (a) and (c) the pervasiveness of × means that the **each** has no effect.

```
      DISPLAY (⊂1 2)×¨⊂3 4
```

```
      DISPLAY (⊂1 2)+.(×¨)⊂3 4
```

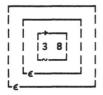

In (b) however the operand of **each** is ∘.× which is *not* a pervasive function. The outer product shape rule forces the final enclosure which is necessary to make the result a scalar

```
      DISPLAY (⊂1 2)∘.×¨⊂3 4
```

In (d) **each** applies to the derived function +.× and effectively cancels one level of enclosure:

DISPLAY B

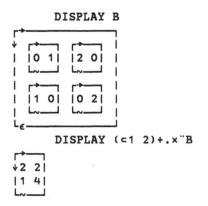

        DISPLAY (⊂1 2)+.×¨B

```
r→──┐
↓2 2|
|1 4|
L~──┘
```

This result is the same as 1 2+.×⊃[1]B for which the steps are:

```
        ⊃[1]B
0 2
1 0

1 0
0 2
        1 2+.×⊃[1]B
2 2
1 4
```

Illustration : Sequences of Inner Products with Nesting

The next two examples demonstrate inner products where one operand is a derived function, and one or both arguments is a nested array.

        DISPLAY (⊂1 2)+.+.×B

```
r→──┐
|3 6|
L~──┘
```

Step 1: the enclosure along the last axis of L gives (⊂⊂1 2), that along the first axis of B gives a two-item vector:

        DISPLAY ⊂[1]B

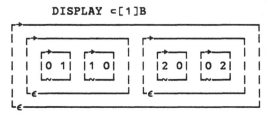

Step 2: form the outer product which the shape rule requires to be a two-item vector:

```
DISPLAY (⊂⊂1 2)∘.×⊂[1]B
```

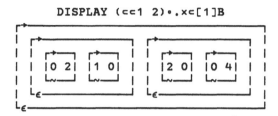

The two items of this outer product are :

```
DISPLAY¨((⊂1 2)×1⊃⊂[1]B)((⊂1 2)×2⊃⊂[1]B)
```

```
r→──────────────────┐    r→──────────────────┐
│ r→───┐ r→───┐      │    │ r→───┐ r→───┐      │
│ │0 2│ │1 0│        │    │ │2 0│ │0 4│        │
│ L~───┘ L~───┘      │    │ L~───┘ L~───┘      │
Lε──────────────────┘    Lε──────────────────┘
```

the derived function +.+ is applied separately to these items:

```
DISPLAY(+.+/(⊂1 2)×1⊃⊂[1]B)(+.+/(⊂1 2)×2⊃⊂[1]B)
```

```
r→───┐
│3 6│
L~───┘
```

or equivalently +.+¨ is applied to the entire outer product:

```
DISPLAY +.+/¨(⊂⊂1 2)∘.×⊂[1]B
```

```
r→───┐
│3 6│
L~───┘
```

The next example differs from the previous one only in the order of execution of the two inner-product operators, and shows what a large difference this can make:

```
DISPLAY (⊂1 2)+.(+.×)B
```

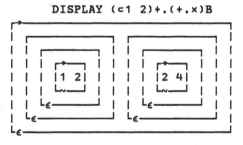

Again, here is a step by step analysis. First construct the outer product (step 2) :

DISPLAY (⊂⊂1 2)∘.(+.×)⊂[1]B

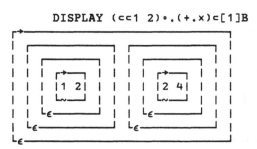

The shape rule shows that this is a two-item vector, whose two items are

DISPLAY (⊂1 2)+.×1⊃⊂[1]B

DISPLAY (⊂1 2)+.×2⊃⊂[1]B

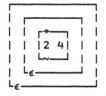

Apply +/ to each item separately:

DISPLAY +/(⊂1 2)+.×1⊃⊂[1]B

DISPLAY +/(⊂1 2)+.×2⊃⊂[1]B

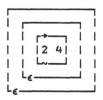

or equivalently apply +/¨ to the vector:

```
     DISPLAY +/"(⊂⊂1 2)∘.(+.×)⊂[1]B
```

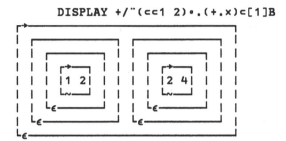

It is rather hard to conceive what the writer of either of the expressions (a) or
(b) might be doing from the application point of view. Nevertheless they demonstrate the care which must be taken in applying rules precisely when coding and
evaluating inner products.

The key message from the above illustrations is that no matter how complex is
an expression which involves inner and outer products its exact meaning and
value can be deduced by careful application of the relevant rules.

5.5.7.1 Inner Product and Scan

There are relationships between scan and inner product in which the four
functions ∧ × ≥ and * play the role of auxiliary functions and triangular
binary matrices form the inner product right argument. For example:

```
     □←UTM←(⍳5)∘.≤⍳5
1 1 1 1 1
0 1 1 1 1
0 0 1 1 1
0 0 0 1 1
0 0 0 0 1

     +\⍳5
1 3 6 10 15
     (⍳5)+.×UTM
1 3 6 10 15

     ×\⍳5
1 2 6 24 120
     (⍳5)×.*UTM
1 2 6 24 120
```

More generally, the relation to be satisfied is

```
     F\A    ↔    A F.aux UTM
```

where A is an a numeric array, UTM is the upper triangular matrix of appropriate shape with 1s on and above the leading diagonal and **aux** is the appro-

priate auxiliary function. The relationship holds for the combinations indicated by the entries y in the tables below:

		aux.					aux.	
		∧	×				≥	⋆
	≠	y	y			=	y	y
	>	y	y			≥	y	y
F	∨	y	y		F	∧	y	y
	+		y			×	y	y
	-		y			≠		y
						⋆	y	y

5.5.7.2 Decode/Encode and Inner/Outer Products

Decode and **encode** share with derived functions the property that they combine the actions of simpler functions, + and × in the case of **decode**, | and ≠ in the case of **encode**. In some special cases there are simple equivalences between **decode/encode** and inner/outer products, for example a polynomial such as $x^2 + 4x + 3$ can be evaluated at $x = 2$ either as

```
(2*2 1 0)+.×1 4 ¯3
```

or as

```
2⊥1 4 ¯3
```

and 13 can be expressed as a binary number either as

```
2|⌊13∘.≠2*3 2 1 0
```

or as

```
(4ρ2)⊤13
```

More interestingly the shape rules for **decode** and **encode** are identical to those for inner and outer products, and the steps for evaluating the inner product of two matrices (see Section 5.5.6) are identical to the first two steps for evaluating an inner product.

Illustration : Decode and encode for arrays

Suppose that

```
(L R)←V53←(2 3ρ0 3 12 10 10 10)(3 2ρ1 5 2 1 7 3)
DISPLAY¨L R
```

```
┌→──────┐  ┌→──┐
↓ 0  3 12│  ↓1  5│
│10 10 10│  │2  1│
└~───────┘  │7  3│
            └~──┘
```

The shape rule gives the shape of L⊥R as 2 by 2. The steps for evaluating L⊥R are:

    Step 1 : Enclose L and R along inner matching axes.
    Step 2 : Perform ⊥ outer product.

```
      L⊥R
 67 195
127 513
```

which is the same as (⊂[2]L)∘.⊥⊂[1]R.

With **encode** the situation is a little more complex. Consider (⍉L)⊤,L⊥R. The shape rule gives the shape of the result as 3 2 4 and the values are:

```
      (⍉L)⊤67 195 127 513
1 5 3 14
0 1 1  5

2 1 1  0
6 9 2  1

7 3 7  9
7 5 7  3
```

The first columns of the result:

```
      1⊃[3](⍉L)⊤67 195 127 513
1 0
2 6
7 7
```

give the separate encodings of **67** with respect to **0 3 12** and **10 10 10**, the second columns the encodings of **195** and so on, while the first and second rows:

```
      1⊃[2](⍉L)⊤67 195 127 513
1 5 3 14
2 1 1  0
7 3 7  9
      2⊃[2](⍉L)⊤67 195 127 513
0 1 1 5
6 9 2 1
7 5 7 3
```

give in their columns the set of codes corresponding to the encoding vectors **0 3 12** and **10 10 10** respectively.

    To reverse the operation L⊥R in the sense of recovering R from each of the two decodings use:

```
      (⊂[2]L)⊤¨⊂[2]L⊥R
 1 5   1 5
 2 1   2 1
 7 3   7 3
```

Exercises 5d

1. **PROD** is a vector of vectors in which alphanumeric characters represent services offered by a set of producers. **CONS** is another vector of vectors describing the various services required by a set of consumers. For example

```
PROD←'ABC'  'BDF'  'AC'  'ABCEF'
CONS←'AB'  'BF'  'ABCD'
```

defines the capabilities of four producers and the requirements of three consumers with regard to a set of six services.

a. Write an expression to give an *incidence matrix* which records which producers can completely supply each consumers requirements, e.g. for the data above the resulting matrix would be

```
1 0 0 1
0 1 0 1
0 0 0 0
```

Amend your code to return a vector of vectors, each of which gives the indices of those producers who can completely satisfy a consumer's requirements, e.g. (1 4)(2 4)(ι0) in the case above.

b. Repeat the above with "partially" replacing "completely" so that the result for the given data is for the data above the resulting matrix would be

```
1 1 1 1
1 1 0 1
1 1 1 1
```

2. a. What does the following phrase do

```
+/¨(⊂10 10)∘.Tι20  ?
```

b. Why does

```
(⊂10 10)+.Tι20
```

give a **DOMAIN ERROR**?

5.6 Applications of User-Defined Operators

5.6.1 Control Structures

The following subsections illustrate how to achieve some traditional control structures of computer science such as *only*, *unless*, *upto* and *until*. The operators ONLY, UNLESS and UPTO provide control based on *arguments*, whereas RPTUNTIL and DOUNTIL give control via successive *results*, either with or without feedback.

5.6.1.1 ONLY

The object of this operator is to execute a function P on the argument L but ONLY on the item which is determined by the index given by Q.

```
[0]    Z+L(P ONLY Q)R        ⍝ P is a function, Q is an index
[1]    Z+L
[1]    (Q⊃Z)+(Q⊃Z)P R
```

Illustration : Selective function application

Actions are performed *only* on array item with a given index.

```
      M
1 2 3
4 5 6
      M+ONLY(1 2)99
1 101 3
4     5 6
      M⌈ONLY(2 3)99
1 2   3
4 5 99
```

5.6.1.2 UNLESS

The operator UNLESS applies a function P to its argument or arguments *unless* a predicate Q is true in which case the right argument is returned.

```
[0]    Z+L(P UNLESS Q)R      ⍝ P is a function, Q is a predicate
[1]    →0 IF Q Z+R           ⍝ exit if predicate true
[2]    →L1 IF 0≠⎕NC 'L'      ⍝ branch if dyadic derived function
[3]    →0 Z+cP R             ⍝ monadic case
[4]    L1:Z+cL P R           ⍝ dyadic case
```

A simple predicate is SOMECHAR:

```
[0]     Z+SOMECHAR R            ⍝ returns 1 if some character items in R
[1]     Z+' '∊∊↑0⍴⊂R
```

Applying UNLESS with **each** allows P to be applied to each item of an array but ignoring any items which satisfy the predicate Q.

```
     V54+(⍳3)(4 5)(⊂'FRED')(7 6)

     2+UNLESS SOMECHAR¨V54
 3 4 5    6 7    FRED    9 8
     ⍋UNLESS SOMECHAR¨V54
 1 2 3    1 2    FRED    2 1
```

The binding rules of Section 5.2 show that the operand Q is tightly bound to UNLESS as the parentheses in the header line suggest visually, that is in the expression 2+UNLESS SOMECHAR¨V it is +UNLESS SOMECHAR to which **each** is applied and *not* SOMECHAR.

Illustration : Selective Processing

Define a nested array containing items of mixed type:

```
     V55+(1 2.5 'XYZ')('ABC' 3 12)
     DISPLAY V55
```

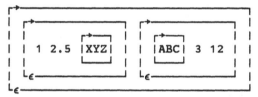

Take the expression 1+V55 and modify it to exclude character items:

```
     1+UNLESS SOMECHAR¨¨V55
 2 3.5 XYZ     ABC 4 13
```

UNLESS can be applied more than once in the same expression:

```
     1(+UNLESS SMALL¨¨)UNLESS SOMECHAR¨V55
 1 2.5 XYZ     ABC 3 13
```

```
[0]     Z+SMALL R        ⍝ returns 1 if first item is less than 5
[1]     Z+5>↑R
```

Explicit parentheses can be used to underline the way in which the two UNLESSs are nested.

```
     1((+UNLESS SMALL)¨ UNLESS SOMECHAR)¨V55
 1 2.5 XYZ     ABC 3 12
```

In this example the SOMECHAR selection must be applied *before* the SMALL selection, otherwise the result would be a DOMAIN ERROR.

Here is a further example in which only integers are selected:

```
      1+UNLESS FRACTNL¨ UNLESS SOMECHAR¨↑V55
2 2.5 XYZ

[0]    Z+FRACTNL R          ⍝ returns 1 if R non-integral
[1]    Z+R≠⌊R
```

5.6.1.3 UPTO

A variation on UNLESS¨ is to stop processing when the predicate is satisfied
rather than skipping an item. A recursive operator to describe this is UPTO.

```
[0]    Z+L(P UPTO Q)R        ⍝ P is a function, Q is a predicate
[1]    →L0 IF Q↑R            ⍝ stop when condition reached
[2]    →L2 IF 0≠⎕NC'L'       ⍝ branch if dyadic derived function
[3]    L1:                   ⍝ monadic case
[4]    →0 Z+(⊂P↑R),(P UPTO Q)1↓R
[5]    L2:                   ⍝ dyadic case
[6]    →0 Z+(⊂L P↑R),L(P UPTO Q)1↓R
[7]    L0:Z+⍳0

      V54+(⍳3)(4 5)(⊂'FRED')(7 6)

      2+UPTO SOMECHAR V54
  3 4 5  6 7
      ⍋UPTO SOMECHAR V54
  1 2 3  1 2
```

The operator UPTO has the general structure

```
[0]    Z+L(P OPR Q)R
[1]    →L0 IF...            ⍝ stopping condition
[2]    →L2 IF 0≠⎕NC'L'      ⍝ branch if dyadic derived function
[3]    →0 Z+...            ⍝ monadic recursion
[4]    L2:→0 Z+...         ⍝ dyadic recursion
[5]    L0:Z+...            ⍝ stopping action
```

UPTO can be abbreviated to two lines by defining a function

```
[0]    Z+LEX
[1]    Z+'L' IF 0≠⎕NC'L'
```

and using execute:

```
[0]    Z+L(P UPTO Q)R        ⍝ P is a function Q a predicate
[1]    →0 IF Q↑Z+R           ⍝ stopping action
[2]    Z+⍎'(⊂',LEX,' P↑R),',LEX,'(P UPTO Q)1↓R'  ⍝ recursion
```

While this has some appeal in packing all the recursive action into one line,
many APL programmers would balk at the obscurity of the code necessary to
do so and would opt for the previous form which also runs faster on account of
the inherent inefficiency of using execute (⍎).

A variation on **UPTO** is to specify a stopping item rather than a predicate. This can be accommodated by a ⎕NC test on Q on entering the function:

```
[0]     Z←L(P Upto Q)R                      ⍝ P is a function
[1]     →L01 IF 3=⎕NC 'Q'                   ⍝ branch if Q is a predicate
[2]     →(L1 IF Q≡↑R),L02                   ⍝ stop condition if value
[3]     L01:→L1 IF Q↑R                      ⍝ stop condition if predicate
[4]     L02:→L2 IF 0≠⎕NC'L'                 ⍝ branch if dyadic
[5]     →0 Z←(⊂P↑R),(P Upto Q)1↓R          ⍝ monadic recursion
[6]     L2:→0 Z←(⊂L P↑R),L(P Upto Q)1↓R    ⍝ dyadic recursion
[7]     L1:Z←⍳0                             ⍝ stopping action

        V54←(⍳3)(4 5)(⊂'FRED')(7 6)

        2↑Upto SOMECHAR V54
  3 4 5 6 7
        2↑Upto(4 5)V54
  3 4 5
```

A programmer intent on shortening code by using **execute** might write:

```
[0]     Z←L(P Upto Q)R                                ⍝ P is a function
[1]     ⍎'→0 IF Q ',('≡' IF 3≠⎕NC'Q'),'↑R',Z←⍳0 ⍝ stopping action
[2]     Z←⍎'(⊂',LEX,' P↑R),',LEX,'(P Upto Q)1↓R'⍝ recursion
```

5.6.1.4 UNTIL

Instead of applying a function P repeatedly to items in the data as in the case of **UNLESS** and **UPTO** (or its variants) it is often desirable to carry on executing P to the *entire* data until some specified circumstance arises. This may or may not involve feedback of the result (cf. the distinction between **POWER1** and **POWER2** in Exercise 5b). Two further distinctions can be made, first is the function monadic or dyadic, and secondly is the test on a predicate or a value. The no feedback case is dealt with by the operator **DOUNTIL**, and feedback by the operator **RPTUNTIL**.

A simple way to develop **DOUNTIL** is to program the monadic case where the test is on a stopping value:

```
[0]     Z←L(P DOUNTIL Q)R;T
[1]     Z←''
[2]     →0 IF Q≡T←P R
[3]     Z←(⊂T),P DOUNTIL Q R

        ?DOUNTIL 3 6                        ⍝ throw a die until a three shows
  5 2 1 5 2 4 5
```

The above example is a further illustration of the application of the *binding rules* (see Section 5.2). The binding between **DOUNTIL** and **3** (right operand binding) is stronger than that between **3** and **6** (vector item binding).

Now extend **DOUNTIL** to deal with the options of dyadic derived function and predicate:

```
[0]    Z←L(P DOUNTIL Q)R;T                    ⍝ P is a function, Q is a test
[1]    Z←''
[2]    →L0 IF 0≠⎕NC 'L'
[3]    →L01 T←P R
[4]    L0:T←L P R
[5]    L01:→L1 IF 3≠⎕NC 'Q'                   ⍝ needs match if Q not predicate
[6]    →0 IF Q T
[7]    →L11
[8]    L1:→0 IF Q≡T
[9]    L11:→L2 IF 0≠⎕NC 'L'
[10]   →0 Z←(⊂T),(P DOUNTIL Q)R
[11]   L2:Z←(⊂T),L(P DOUNTIL Q)R
```

```
[0]    Z←L ROLL R
[1]    Z←?L⍴R
```

```
[0]    Z←ALIKE R
[1]    Z←∧/R=↑R
```

```
       2 ROLL DOUNTIL ALIKE 6                 ⍝ throw a pair of dice until
 6 5   5 2   2 4   6 3                         ⍝ a double appears
```

For the purposes of copy-typing the condensed form using **execute** (⍎) is often more useful:

```
[0]    Z←L(P DOUNTIL Q)R;T
[1]    T←⍎LEX,' P R',Z←''
[2]    ⍎'→0 IF Q ',('≡' IF 3≠⎕NC'Q'),'T'
[3]    Z←(⊂T),⍎LEX,'(P DOUNTIL Q)R'
```

The function RPTUNTIL is developed in the same way, that is, first by defining one case, e.g. where the function P is dyadic and Q is a predicate:

```
[0]    Z←L(P RPTUNTIL Q)R
[1]    →0 IF Q Z←R
[2]    Z←L(P RPTUNTIL Q)L P R'
```

and then generalizing it to cover the other cases by using ⍎:

```
[0]    Z←L(P RPTUNTIL Q)R                     ⍝ P is a function, Q is a test
[1]    ⍎'→0 IF Q ',('≡' IF 3≠⎕NC'Q'),'Z←R'
[2]    Z←⍎LEX,'(P RPTUNTIL Q)',LEX,' P R'
```

```
[0]    Z←SMALL R                              ⍝ returns 1 if first item less than 5
[1]    Z←5>↑R
```

```
       2↓RPTUNTIL SMALL ⌽⍳13
3 2 1
       .5×RPTUNTIL SMALL 999
3.902
       'X',RPTUNTIL 'XXXX' ''
XXXX
```

Illustration : Repetitive Prompts

The simplest way of combining N separate input strings from a terminal into an N-item vector is

```
    ♠¨Nρ'⎕'
```

Entry of multiple input lines is usually associated with prompts and a simple function which provides these is

```
[0]    Z←ASK R
[1]    ⎕←R
[2]    Z←(ρ,R)↓⎕
```

```
       ASK 'NAME='
NAME=ALF
ALF
```

ASK can be used with **each** to obtain answers to an ordered succession of prompts:

```
       ASK¨'NAME='  'NO='
NAME=ALF
NO=49
 ALF 49
```

Now define a function NULL (that is null line) for use as a stopping condition:

```
[0]    Z←NULL R
[1]    Z←0∊ρR
```

and apply DOUNTIL to issue repeated prompts:

```
       ASK DOUNTIL NULL 'ENTER='
ENTER=A
ENTER=BB
ENTER=CDE
ENTER=
 A BB CDE
```

```
       ASK DOUNTIL '99' 'ENTER='
ENTER=7
ENTER=33
ENTER=99
 7 33
```

The derived function ASK¨ can be used in conjunction with DOUNTIL to repeat *chains* of prompts until a complete cycle of null responses has been given:

```
      ASK"DOUNTIL NULL 'NAME=' 'NO='
NAME=ABC
NO=1
NAME=XYZ
NO=2
NAME=
NO=
   ABC 1    XYZ 2
```

Illustration : Iterative solution of non-linear equations

RPTUNTIL provides a method of solving by iteration equations which can be
expressed in the form $y = f(y)$ for which there is a convergent solution from the
given start value.   An example is $y = \cos(y)$ which was first discussed in
Exercise 5b4c.   EPS is a global variable which defines a stopping tolerance, e.g.
.00001 in the present case.

```
[0]    Z+COS X
[1]    Z+2oX

[0]    Z+NEAR X
[1]    ⍝ n.b. function P is defined in RPTUNTIL
[2]    Z+EPS>|X-P X

       COS RPTUNTIL NEAR 1
0.73909
```

For Newton-Raphson iteration define another two operators and a variable containing the step size:

```
[0]    Z+(P NEWTON)X
[1]    Z+X-(P X)÷P DERIV X

[0]    Z+(P DERIV)X
[1]    Z+((P X+ΔX)-P X)÷ΔX

       ΔX+.00005
```

To solve the equation $x(x-1) = 2$ define

```
[0]    Z+F X
[1]    Z+2-X×X-1
```

The roots to which the Newton-Raphson process converges for different starting
values are then given by:

```
       (F NEWTON)RPTUNTIL NEAR 1
2
       (F NEWTON)RPTUNTIL NEAR ¯1.2
¯1
```

Illustration : Non-linear function fitting

The primitive function ⊞ performs least squares fits of linear functions. With only a modest amount of programming it can also be used to fit a much wider range of non-linear functions as the present illustration shows. Suppose that it is required to fit a function of the form

$$y = a + b.exp(-cx)$$

to the data

```
      X←V56
0 1 2 3 4 5 6 7 8 9 10 11 12 13 14 15 16 17 18 19
      Y←V57
4.745 4.6532 4.6036 4.0066 4.0864 4.5687 3.806 3.1908
      3.0976 3.6759 3.8764 3.4329 4.1062 3.0066 2.6309
      3.6943 3.2929 2.4183 3.4453 3.1949
```

Define the function FN in which C stands for coefficients:

```
[0]    Z←C FN X
[1]    Z←C[1]+C[2]×*-C[3]×X
```

Partial derivatives with regard to the coefficients are estimated by defining an operator:

```
[0]    Z←(C(F PDERIV X)N;T
[1]    ⍝ N is index of coefficient whose partial derivative is required
[2]    Z←(((C+T×N=⍳⍴C)F X)-C F X)÷T←ΔX3[N]
```

Intervals can be defined for each coefficient separately as the rightmost part of the above function line implies:

```
      ΔX3
0.00001 0.00001 0.00001
```

Fitting is carried out using domino:

```
[0]    Z←X(F FIT Y)C
[1]    Z←C+(Y-C F X)⊞[1](⊂C)(F PDERIV X)¨⍳⍴C
```

Make a first guess at the coefficients:

```
      C0←3 4 .4
```

and then run the function FIT:

```
      X FN FIT Y C0
3.2267 1.4653 0.21462
```

Now use RPTUNTIL to iterate towards a solution with prescribed accuracy. A stopping criterion might be that all the coefficients are within EPS of the previous iteration. This is described by the function ALLNEAR which uses the P and L of RPTUNTIL:

```
[0]    Z←ALLNEAR C
[1]    ⍝ n.b. function P and argument L are defined in RPTUNTIL
[2]    Z←∧/EPS>|C-C P L
```

```
       EPS
0.00001
```

A small amendment must be made to FIT:

```
[0]    Z←A(F Fit Y)B
[1]    Z←R+(Y-R F L)⌹⊃[1](⊂R)(F PDERIV L)¨⍳⍴R
```

which now has dummy arguments A and B since it is the derived function
(F Fit Y) which is the left operand of RPTUNTIL and L and R are the arguments of RPTUNTIL. The iterated solution is:

```
       X FN Fit Y RPTUNTIL ALLNEAR 3 4 .4
2.8907 1.9388 0.1186
```

To obtain a trace of the steps towards convergence add ⎕← at an appropriate
point in RPTUNTIL:

```
[2]    Z←⍺LEX,'(P RPTUNTIL Q)⎕←',LEX,' P R'
```

and rerun:

```
       X FN Fit Y RPTUNTIL ALLNEAR 3 4 .4
3.2267 1.4653 0.21462
3.0237 1.7732 0.077427
3.0959 1.7285 0.12798
2.8934 1.9353 0.11731
2.8909 1.9387 0.11862
2.8907 1.9388 0.1186
```

To confirm the correctness of Fit define Y so that the exact result is known in
advance:

```
       Y←5+2×⋆-.2×X
       X FN Fit Y RPTUNTIL ALLNEAR 3 4 .4
3 4 0.4
5.1016 1.8463 0.2698
5.0267 1.9624 0.18631
5.0021 1.9973 0.20003
5 2 0.2
```

5.6.2 Conditional and Alternative Function Execution

Operators are a natural mechanism for writing functions which avoid
anticipatable APL errors. The first illustration below gives an operator which
restricts function execution to selected parts of the data only. The second illustration gives a technique for providing alternative monadic functions.

Illustration : Data Filtering

The operator UNLESS in Section 5.6.1.2 was used to to apply a left operand
selectively. A disadvantage of this is that the predicate function may generate an
error as in:

```
V55←(1 2.5 'XYZ')('ABC' 3 12)
DISPLAY V55
```

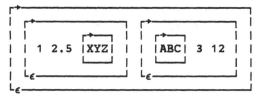

```
[0]    Z←SMALL R
[1]    Z←5>↑R
```

```
       1+UNLESS SMALL¨V55
DOMAIN ERROR
SMALL[1]
[1]    Z←5>↑R
```

A technique which overcomes this is to to apply **compression** to R in order that P
be applied only to those items which satisfy a predicate Q such as NUM:

```
[0]    Z←NUM R                        ⍝ returns 1 if R entirely numeric
[1]    Z←2=2⊥(0 ' ')∊∊↑0ρ⊂R
```

Next an operator FILTER is defined:

```
[0]    Z←L(P FILTER Q)R;T             ⍝ P is a function, Q a predicate
[1]    Z←↑R((T/R)←L P(T←Q¨R)/R)
```

```
       1+FILTER NUM¨¨V55
  2 3.5 XYZ      ABC 3 13
       1(+FILTER SMALL¨)FILTER NUM¨¨V55
  2 3.5 XYZ      ABC 4 12
```

Another possible filter function is INTEGRAL:

```
[0]    Z←INTEGRAL R                   ⍝ returns 1 if R is numeric and integral
[1]    Z←R≡⌊R                         ⍝ n.b. must be **match**, = won't do!
```

```
       1(+FILTER INTEGRAL¨)FILTER NUM¨¨V55
  2 2.5 XYZ      ABC 4 13
```

Illustration : The ELSE Operator

It can sometimes be convenient to be able to provide alternative monadic functions depending on some user-defined condition which need not necessarily be error-producing. This facility is provided by the operator ELSE:

```
[0]    Z←L(P ELSE Q)R          ⍝ P and Q are functions
[1]    →L1 IF~L                ⍝ Q is executed if L false
[2]    →0 Z←P R                ⍝ otherwise P is executed
[3]    L1:Z←Q R

       (A≠2)(÷ELSE+)A←2 3 4
2 3 4
       (A≠2)(÷ELSE+)¨A←2 3 4
2 0.3333 0.25
```

The rightmost sets of brackets are not necessary in the above :

```
       (A≠2)÷ELSE+¨A←2 3 4
2 0.3333 0.25
```

but they help clarify the meaning.

Exercises 5e

1. Write an operator ONLYS which extends ONLY by recursion so that Q is a *vector* of indices, e.g.

```
       M←2 3⍴⍳6
       M+ONLYS((2 2)(2 3))100
1   2   3
4 105 106
```

2. a. Rewrite the operator TRACE of Section 5.3 using ⍛ and the function LEX.

   b. Extend the operator SIMPLE in Section 5.4 so that it deals with both monadic and dyadic derived functions, e.g.

```
       ⍋Simple((9 5 6)(7 4))(8 5 3)
```

should return the value ((2 3 1)(2 1))(3 2 1).

3. Rewrite the dyadic composition operator COMP1 of Section 5.2.2 so that it deals with both monadic and dyadic derived functions, that is ⍴Comp1⍴¨T is equivalent to ⍴¨⍴¨T, and 2∈Comp1∈¨T is equivalent to 2∈¨∈¨T.

4. An alternative to the Newton-Raphson technique for finding a root of $f(x) = 0$ is known as the Secant method. The algorithm consists of starting with a pair of values x_0 and x_1 one on each side of the root, and identifying the co-ordinate, x_2, of the point where the line joining the points $(x_0, f(x_0))$ and $(x_1, f(x_1))$ crosses the x-axis. Take whichever of the intervals (x_0, x_2) and (x_2, x_1) contains the root, and repeat. Under suitable conditions x_2 will converge to the root.

Using **RPTUNTIL** and **NEAR** write an operator analogous to **NEWTON** together with any requisite operators or functions so that

    F SECANT RPTUNTIL NEAR V

where **V** is an appropriate two-item vector of co-ordinates, delivers the required root of **F**.

5.6.3 LEVEL

In first-generation APL, the result of adding two vectors of equal length is unambiguous:

```
      1 2+1 3
2 5
```

Nesting allows two other possibilities for adding simple vectors.

```
(a)    1        2        (b)    1        3
       |        |               |        |
      1——3    1——3            1——2    1——2
```

Assume that an operator LEVEL has been written to distinguish these cases: In (a) the vector 1 3 is nested (i.e. at level 1) and is added to each item of vector 1 2:

```
      1 2+LEVEL(0 1)1 3
2 4   3 5
```

In (b) the vector 1 2 is nested (level 1) and added to each item of vector 1 3 :

```
      1 2+LEVEL(1 0)1 3
2 3   4 5
```

The operator LEVEL might be written

```
[0]    Z+L(P LEVEL Q)R      ⍝ P is a function, Q indicates depth
[1]    →L1 IF 0≠⎕NC'L'       ⍝ branch if dyadic
[2]    →0 Z+(P MONLEV Q)R    ⍝ P monadic
[3]    L1:Z+L(P DYALEV Q)R   ⍝ P dyadic
```

where MONLEV and DYALEV deal with the monadic and dyadic cases respectively.

```
[0]    Z+(P MONLEV Q)R      ⍝ Q is a non-negative integer
[1]    →L1 IF Q≥≡R
[2]    →0 Z+P MONLEV Q¨R
[3]    L1:→0 Z+P R
```

In line 1 of DYALEV, the requested depths Q are compared with the actual depths of the arguments L and R - if either is lower a level of nesting is removed by **each** with, if necessary, an **enclose** of the other.

```
[0]    Z+L(P DYALEV Q)R      ⍝ Q a 2-item vector of non-negative integers
[1]    →(2⍳Q<≡¨L R)CASE(0,L00)(1,L01)(2,L10)(3,L11)
[2]    L00:→0 Z+L P R                ⍝ depth reached for both L and R
[3]    L01:→0 Z+(⊂L)P DYALEV Q¨R     ⍝ depth reached for L but not R
[4]    L10:→0 Z+L P DYALEV Q¨⊂R      ⍝ depth reached for R but not L
[5]    L11:Z+L P DYALEV Q¨R          ⍝ depth reached for neither L nor R

[0]    Z+L CASE R           ⍝ L is an expression, R a vector of 2-item
[1]    Z+(L≡¨↑¨R)/2⊃¨R       ⍝   vectors, each is (value of L, label)
```

The next two subsections use the same nested vector T. T is used as its name rather than V in order to emphasize that it is to be thought of as modelling a *tree*.

```
    ≡T←(6(8 3))(1 2)
3
```

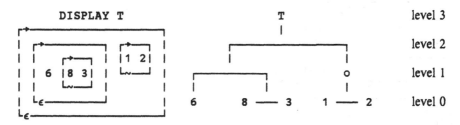

Since every subtree extends to level 0, it is necessary to insert extra levels in the case of non-uniform trees.

5.6.3.1 LEVEL with Monadic P

In the following illustrations the function Φ is taken as P.

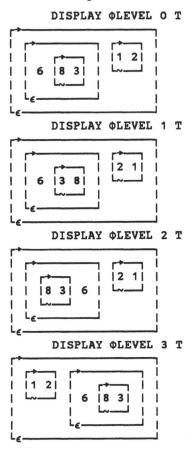

5.6.3.2 LEVEL with Dyadic P :

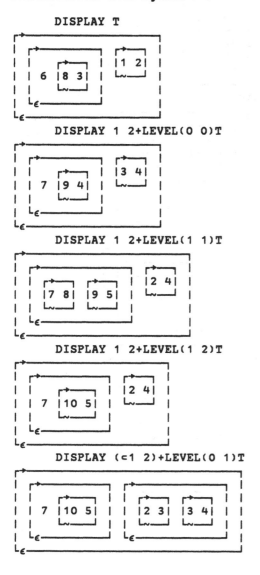

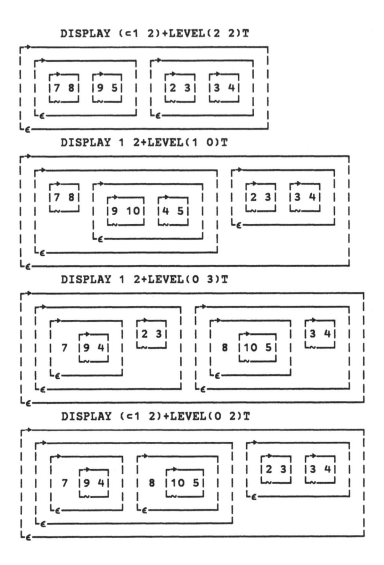

Exercises 5f

The following exercises assume that a *name* is a vector of character vectors, e.g.

   NAME←'JAMES' 'ANTHONY' 'LAMB'

and that NAMES is a vector of eight names, viz:

```
        ⊃NAMES
JAMES    ANTHONY    LAMB
HUGH     WILLIAM    JONES
FRED     SMITH
ARTHUR   WILLIAM    DALY
HAMISH   MCGREGOR
ANDREW   DAVID      WILLIAM   MASON
SEAN     EWAN       MCTAVISH
ANDREW   WILLIAM    MASON
```

1. For a single name, e.g. **NAME** define a function **SHORTEN** which replaces the forenames with a vector of initials so that the name becomes a two-item vector e.g.

```
      SHORTEN NAME
JA    LAMB
```

2. Define a predicate **ISW2** which returns 1 if the second name is **WILLIAM** and another predicate **SCOTCH** which returns 1 if the first two characters of the surname are **MC**.

3. Print a matrix with one row per name which contains everyone's names except those whose second name is **WILLIAM**.

4. Print a similar matrix of names in which all except Scotsmen have their forenames abbreviated to their initials.

5. Define a function **LENGTHEN** to allow the printing of a matrix of everyone's names in which Scotsmen have the first two characters of their surnames replaced by **MAC**.

6. Print a matrix of names with one row per name in which all surnames are aligned.

Summary of Functions used in Chapter 5

Section 5.2
HTD   convert hexadecimal to decimal
DTH   convert decimal to hexadecimal

Exercises 5a
FROMDEC   convert to decimal from arbitrary number base
TODEC   convert to arbitrary number base from decimal

Section 5.4
PATH   find path to given item in a vector
ENLIST   selective enlist

Exercises 5b
PRODUCT   description of ×/
JOIN   description of ,/
CHALL   change all occurrences
CODIFY   encrypt a character string
MEAN   mean of a numeric vector
MEDIAN   median of a numeric vector

Section 5.5.3
SPIRAL   co-ordinates of spiral in Euclidean plane

Section 5.5.4
EXPAND   function form of expand operator

Section 5.5.6
AD   absolute difference
GRAD   gradient of line in Euclidean co-ordinates
MIDPT   midpoint of line in Euclidean co-ordinates

Exercises 5c
SUBMAT   returns all submatrices of a given shape in matrix
PATIN   matches a binary pattern in a binary matrix

Section 5.6.1.2
SOMECHAR   returns 1 if some character items in right argument
FRACTNL   returns 1 if right argument non-integral

Section 5.6.1.3
LEX   auxiliary function for writing ambi-valent operators

Section 5.6.1.4
ALIKE   returns 1 if all items in vector equal
ASK   returns answer following prompt
NULL   predicate is-null-vector

COS	cosine
NEAR	auxiliary function to provide stop for operator RPTUNTIL
ALLNEAR	generalization of function NEAR
NUM	returns 1 if all items in array numeric

Section 5.6.2

INTEGRAL	returns 1 if array numeric and all items integers
DECODE	deciphers an encrypted character string

Section 5.6.3

CASE	case statement

Exercises 5f

SHORTEN	replaces names with initials
ISW2	predicate is-second-name WILLIAM
SCOTCH	predicate is-first-two-characters MC
LENGTHEN	replaces MC with MAC

Summary of User-defined Operators in Chapter 5

Section 5.1

COM   commutes arguments
SEE   dynamic trace (monadic functions)
ALONG   progresses functions left to right
TABLE   outer product of vector with itself

Section 5.2

RED   reduction along axis
LRED   reduction from left along axis
HEX   hexadecimal arithmetic
HEXE   HEX each
NEXT   joins matrices of non-compatible shapes
CONSEC   indices of start points of sequences

Exercises 5a

BASE   arithmetic in arbitrary base
ROOTOP   pth. root

Section 5.3

COMP1   function composition : L P Q R
COMP2   function composition : P L Q R
TRACE   dynamic trace (ambi-valent functions)

Section 5.4

SIMPLE   makes non-pervasive functions penetrate to simple items

Exercises 5b

POWER1   function to the power : (L P) repeated Q times starting with R
POWER2   function to the power : (P R) repeated Q times starting with L
POLISH   polishes a matrix by subtracting from rows and columns

Section 5.5

UNSCAN   reverses scan using R P L
UNDO   reverses scan using L P R

Section 5.6

ONLY   function executed only for given index
ONLYS   function executed only for given indices
UNLESS   function applied to items unless they satisfy predicate
UPTO   function applied item by item until value found in argument
DOUNTIL   function applied without feedback
RPTUNTIL   function applied with feedback
NEWTON   root-finding by Newton-Raphson iteration
DERIV   derivative of function

PDERIV   partial derivatives of function coefficients
FIT   operator for fitting non-linear functions
FILTER   processes data only if predicate true
ELSE   alternative monadic functions dependent on user condition
LEVEL   applies function at given depth levels of argument(s)
MONLEV   monadic case of **LEVEL**
DYALEV   dyadic case of **LEVEL**

6
Advanced Modelling and Data Structures

Chapter 3 interrupted the discussion of APL features in order to give some simple applications in which nested arrays prove their worth. This chapter performs a similar role, but the applications are now typical problems which arise in Operational Research and involve more sophisticated uses of the techniques of the previous chapters. The problem situations involved are designed to illustrate how some commonly occurring data structures can be modelled in APL2, and how APL2 programs can be built round them in a systematic fashion.

6.1 Trees Without Keys

There are many different ways to build tree-like data structures of which three will be discussed in the first three sections of this chapter. The simplest sort of tree is a hierarchical one in which nested arrays are used to model subordinate relationships, e.g. the organization of a department:

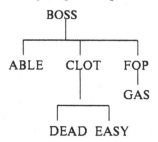

is described by the following nested array:

```
HT←'BOSS'('ABLE' 'CLOT'('DEAD' 'EASY') 'FOP'(⊂'GAS'))
DISPLAY HT
```

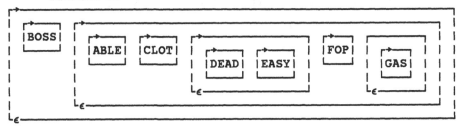

Notice that it is necessary to enclose the character string when a member has only one subordinate.

A straightforward enlist is of little value ...

```
    ∈HT
BOSSABLECLOTDEADEASYFOPGAS
```

... however the selective enlist function defined in Section 5.4 can give us an overall namelist ...

```
[0]    Z←L ENLIST R
[1]    →L1 IF L≥≡R
[2]    →0 Z←↑,/L ENLIST¨R
[3]    L1:Z←⊂R
```

```
    DISPLAY 1 ENLIST HT
```

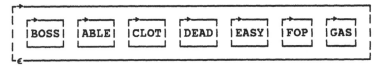

... or select members at either of the top two levels:

```
    DISPLAY 2 ENLIST HT
```

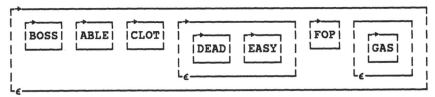

6.2 Trees with Keys

Another way in which a tree structure can be used to store data in a way which reflects its internal relationships involves using keys.

Suppose a hierarchical data set is structured as shown in the diagram below in which numbers represent keys, "*" represents a subtree which is expanded at the next lower level and letters of the alphabet represent data which might in practice be very large.

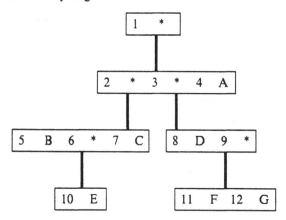

A tree is thus a vector with an even number of items - odd numbered items are the unique numeric keys, even numbered items are either subtrees or data items. The functions to be discussed concern the *structure* of and navigation through trees and are entirely independent of the data which the trees are used to store. The tree sketched above could be modelled as

$$\text{TREE}\leftarrow 1(2(5'B'6(10'E')7'C')3(8'D'9(11'F'12'G'))4'A')$$

It is worth spending a few moments to accustom oneself to the relationship between the drawn form of the tree and the DISPLAYed version:

DISPLAY TREE

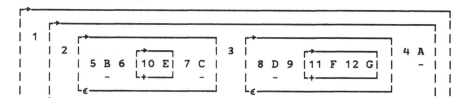

Consider the problem of finding the path to a given key L. The nature of this tree structure guarantees that it is not necessary to provide for "level-breakers" (i.e. empty vectors - see Section 1.3.1). If it is further assumed that L must be present in R and that the key values do not occur in the data then the initial PATH algorithm of Section 5.4 is sufficient.

```
[0]     Z+L PATH R;T          ⍝ L a numeric scalar, R a tree
[1]     →L1 IF 1≤≡R           ⍝ branch if R nested
[2]     →0 Z←⍳0
[3]     L1:T←(L∊¨∊¨R)⍳1        ⍝ identify subtree T at current depth ..
[4]     Z←T,L PATH T⊃R         ⍝ .. then find path within subtree T

        7 PATH TREE
2 2 5
```

Several paths may be found in one action:

```
        DISPLAY (8 9 10)PATH¨⊂TREE
```

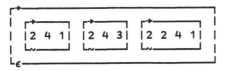

The function **PATH** is the inverse of **pick**:

```
        (8 9 10 PATH¨⊂TREE)⊃¨⊂TREE
8 9 10
```

6.2.1 Finding Ancestors

Define the *ancestors* of a key as those keys which precede it in a path. For a given key L the function **ANCIN** (short for "Ancestor in") which has a very similar structure to **PATH** provides a trace of all ancestors from a given key.

```
[0]     Z←L ANCIN R;T
[1]     →L1 IF~L∊R
[2]     →0 Z←⍳0
[3]     L1:T←(L∊¨∊¨R)⍳1        ⍝ identify subtree T at current depth ..
[4]     Z←R[T-1],L ANCIN T⊃R   ⍝ .. then find ancestor within subtree T

        8 ANCIN TREE
1 3
        10 ANCIN TREE
1 2 6
```

```
        DISPLAY (⍳10)ANCIN¨⊂TREE
```

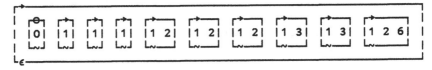

6.2.2 Subtrees

A subtree can be defined as the portion of a tree which is identified by a key. A subtree is therefore completely determined by the path to its key. Adding one to the lowest order item of the path produced by the **PATH** function yields the path to the subtree itself. The function **STPATH** produces the path to the subtree:

```
[0]    Z+L STPATH R
[1]    Z+((-ρZ)↑1)+Z+L PATH R
```

from which the subtree itself may be obtained using **pick**:

```
[0]    Z+L SUBT R
[1]    Z+(L STPATH R)⊃R
```

Hence subtrees can be exhibited:

          DISPLAY 2 SUBT TREE

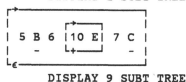

          DISPLAY 9 SUBT TREE

```
┌→──────────────┐
│11  F  12  G│
└+──────────────┘
```

6.2.3 Eliminating and Swapping Subtrees

The function **CUTFROM** removes the subtree associated with key **L**.

```
[0]    Z+L CUTFROM R
[1]    Z+↑R((L STPATH R)⊃R)+⊂ι0
```

          DISPLAY 2 CUTFROM TREE

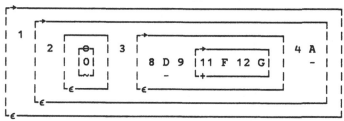

SWAP exchanges the subtrees associated with keys **L[2]** and **L[1]**:

```
[0]    Z←L SWAP R;T
[1]    Z←R
[2]    T←L STPATH¨⊂Z
[3]    Z←((⌽T)⊃¨⊂Z)←T⊃¨⊂Z
```

       DISPLAY 7 3 SWAP TREE

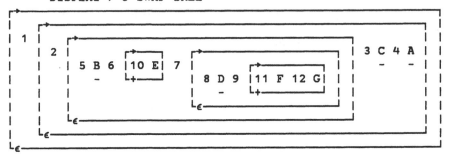

6.3 Binary Trees

This section discusses another form of tree structure in which there are no keys but instead the structure depends on properties of the data, together with the order in which it is entered into the tree. A tree consists of three components, a root, a left subtree, and a right subtree. Data resides only at the roots of trees and subtrees, and as in the previous section it can always be made into a scalar however complex the *actual* structure of the data which resides there. The left and right subtrees may be empty, and it is natural to represent empty trees as ι0. Subtrees repeat the structure of trees so that a typical three-level tree is:

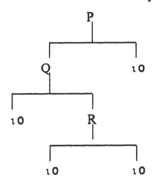

A simple binary tree operation involves storing names which are ordered alphabetically The first name is entered at the root, the next name goes to the left or right depending on whether it comes before or after the root word in the alphabet. Subsequent names enter at the root and traverse the tree going left or right at every subtree root until an empty subtree is found where the incoming name can be inserted. Thus if FRED, ANNE and DAVID are to be entered in that order the resulting tree is

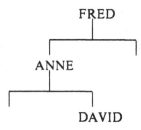

If they are entered in the reverse order the tree is

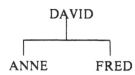

Basic operations on binary trees are *get-root* (modelled by 2⊃TREE), *get-left-subtree* (↑TREE), *get-right-subtree* (3⊃TREE), and *is-empty* (0=ρTREE). At a secondary function level, useful functions are those which carry out operations like *insert*, *search*, *make-tree*, *count-leaves*, *count-comparisons* and *is-equivalent*.

Since binary trees are recursively defined, it is not surprising to find that recursive functions are a natural way of building up secondary functions.

First consider binary trees which have simple numeric scalars as the root values. The function INS implements an algorithm for inserting an item into a tree which uses "multi-level recursion," that is INS calls ΔINS which calls ΔΔINS which calls INS.

To insert an item L into a tree R first test to see whether R is empty, in which case the result is (ι0)L(ι0):

```
[0]    Z←L INS R          ⍝ L is item, R is tree
[1]    →L1 IF 0=ρR        ⍝ return (ι0)L(ι0) if tree empty
[2]    →0 Z←L ΔINS R      ⍝ else ...
[3]    L1:Z←R L R
```

ΔINS deals with the non-empty case and tests first whether L matches the root, in which case there is nothing to do.

```
[0]    Z←L ΔINS R
[1]    →L1 IF L≡2⊃R       ⍝ return tree unchanged if L matches root
[2]    →0 Z←L ΔΔINS R     ⍝ else ...
[3]    L1:Z←R
```

ΔΔINS deals with the general case where L does not match the root in which case L is inserted on the left or right depending on whether it is less than or greater than the root:

```
[0]    Z←L ΔΔINS R
[1]    →L1 IF L>2⊃R       ⍝ go right if L > root
[2]    →0 Z←(⊂L INS↑R),1↓R ⍝ else go left
[3]    L1:Z←(2↑R),⊂L INS↑⌽R
```

The **INS** function also allows a binary tree to be constructed from scratch by successive insertions using a right argument of ιOs. As in the previous section it is worth spending a few moments observing the relationship between the tree diagrams given above and the **DISPLAY**ed versions.

```
TR1←6 INS ιO
DISPLAY TR1
```

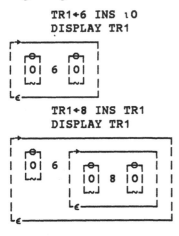

```
TR1←8 INS TR1
DISPLAY TR1
```

```
TR1←7.5 INS 9 INS TR1
DISPLAY TR1
```

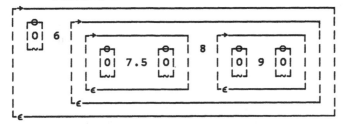

The above code suggests a further recursive function which converts a vector into a binary tree:

```
[0]    Z←MAKET R
[1]    →L1 IF 0=ρ,R         ⍝ if empty argument return empty tree
[2]    →0 Z←(↑R)INS MAKET 1↓R   ⍝ else insert first into tree
                            ⍝         made from remainder
[3]    L1:Z←ιO
```

so that

```
TR1←MAKET 7.5 9 8 6
```

also constructs the tree shown in the last example.

6.3.1 Trees with non-simple Scalar Nodes

One of the strengths of APL2 is that if the underlying structure is now changed to one of arbitrary complexity the upgraded code for the INS and ISIN function sequences is virtually unaltered. To be specific, suppose that the node items are identified by keys which are taken to be the first item of nested vectors such as V61, V62, and V63 below.

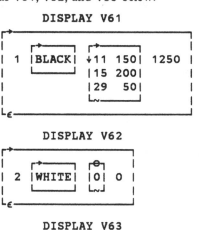

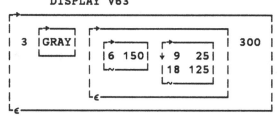

The only change necessary is to the function ∆∆INS where **firsts** must be added to the conditional clause:

```
[1]    →L1 IF (↑L)>↑2⊃R
```

A tree to store the three nested arrays P, Q and R is constructed by:

```
    TR2←MAKET V61 V62 V63
```

The resulting tree TR2 has the shape

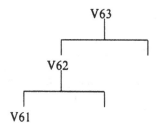

6.3.2 Searching Binary Trees

A function sequence which tests whether or not an item is present in a tree has
an almost identical recursive structure to that of the functions INS, ΔINS, and
ΔΔINS used for inserting items into a tree.

```
[0]   Z←L ISIN R            ⍝ L is item, R is tree
[1]   →L1 IF 0=⍴R           ⍝ return ⍳0 for empty tree
[2]   →0 Z←L ΔISIN R        ⍝ else ...
[3]   L1:Z←0

[0]   Z←L ΔISIN R
[1]   →L1 IF L≡2⊃R          ⍝ see if root matches
[2]   →0 Z←L ΔΔISIN R       ⍝ else ...
[3]   L1:Z←1

[0]   Z←L ΔΔISIN R
[1]   →L1 IF (↑L)>↑2⊃R      ⍝ try right if L>root
[2]   →0 Z←L ISIN↑R         ⍝ else go left
[3]   L1:Z←L ISIN↑⌽R
```

```
      7.5 ISIN TR1
1
      7 8 9 ISIN¨⊂TR1
0 1 1
      (V61 V62 15)ISIN¨⊂TR2
1 1 0
```

Of course, if the data items are simple numeric scalars, ISIN can be achieved
much more simply by L∊R, e.g.

```
      7 8 9∊∊TR1
0 1 1
```

Depth-first scan, that is a traverse of the tree which penetrates each path as
deeply as possible into the tree before retreating and fanning out to other nodes,
is also achieved trivially through enlist:

```
      ∊TR1
6 7.5 8 9
```

and the number of leaves in a tree by

```
      ⍴∊TR1
4
```

6.3.3 Selective Enlist with Binary Trees

Walking the tree TR2 in a depth-first fashion poses a problem, because **enlist** is
once again too heavy-handed for the job, and steam rollers everything down to
scalar level:

```
     ∈TR2
1 BLACK 11 150 15 200 29 50 1250 2 WHITE 0 3 GRAY 6 150 9 25
18 125 300
     ρ∈TR2
32
```

The selective enlist function **ENLIST** defined in Section 5.4 provides the ability to scan the items in **TR2** in depth-first fashion while retaining the simple (i.e. non-nested) structures, namely character and numeric arrays:

DISPLAY 8↑1 ENLIST TR2

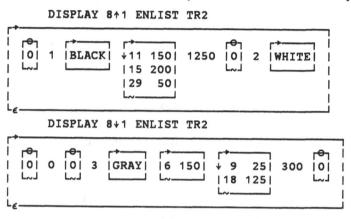

DISPLAY 8↓1 ENLIST TR2

6.3.4 Data-equivalent Binary Trees

Different binary trees are data-equivalent if they contain the same data but in a different tree structure. The differences arise only on account of the items being inserted in different orders. For example:

DISPLAY MAKET 4 5 6

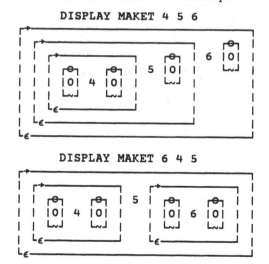

DISPLAY MAKET 6 4 5

whose corresponding tree structures are

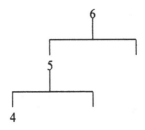

and

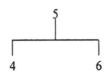

The (ι 0)s representing empty left and right sub-trees are generated by line
[3] of MAKET.

Simple **enlist** is adequate to compare trees for data-equivalence, for example:

```
(εMAKET 4 5 6)≡εMAKET 6 4 5
1
```

This test applies equally to trees with complex underlying structure:

```
(εMAKET V61 V62 V63)≡εMAKET V62 V63 V61
1
```

6.3.5 Alternative Comparisons

If the root items are not numeric scalars it is necessary to replace > in ΔΔINS
and ΔΔISIN with a function GT which determines in context an appropriate
definition of "is greater than." For example, if the root items are character-
string vectors an appropriate GT function which exploits dyadic **grade-up** is

```
[0]    Z←L GT R
[1]    →L1 IF 0≠↑0⍴L      ⍝ use collating sequence if L non-numeric
[2]    →0 Z←(↑L)>↑R       ⍝ else use >
[3]    L1:Z←>/⎕AV⍋⊃L R    ⍝ test for L before R in alphabetic order
```

Only some relatively small details of the ΔΔINS and ΔΔISIN functions need
be changed. If the right argument of MAKET has only one name, this must be
enclosed.

```
[0]    Z←L ΔΔIns R
[1]    →L1 IF(↑L)GT 2⊃R
[2]    →0 Z←(⊂L Ins↑R),1↓R
[3]    L1:Z←(2↑R),⊂L Ins↑⌽R

[0]    Z←L ΔΔIsin R
[1]    →L1 IF(↑L)GT 2⊃R
[2]    →0 Z←L Isin↑R
[3]    L1:Z←L Isin↑⌽R
```

These functions together with `Ins`, `Isin`, `ΔIns`, `ΔIsin`, and **`Maket`** suitably adapted cover all the three types of data which have been considered.

```
TR3←'FRED' Ins ι0
DISPLAY TR3
```

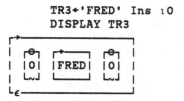

```
DISPLAY TR3←Maket 'DAVID' 'ANN' 'FRED'
```

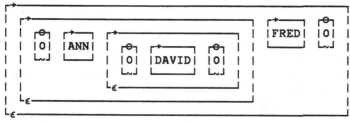

```
      'ANN' 'JANE' Isin¨⊂TR3
1 0
      ∈TR3
ANNDAVIDFRED
```

Exercises 6a

1. Amend the `Isin` sequence of functions so that `Isin` counts the number of comparisons which are made in searching for `L`.

2. Write a function sequence `SUB`, `ΔSUB`, `ΔΔSUB`, similar to the `ISIN` sequence, which obtains the subtree in `R` at node `L`.

6.4 Networks

This section considers closed structures with arcs and nodes. Two nodes can be identified as source and sink respectively, and the other nodes are intermediate nodes. This structure typically models a situation where something like a fluid or an electric current flows from the source to the sink. Flow may only occur in one direction along each arc and the arcs also have capacity constraints, that is each has a maximum flow which it can sustain. A frequent objective is to maximize the total overall flow from source to sink, so that for example the maximum amount of oil is conveyed through a system of pipes from the oil well to the refinery.

The so-called PERT diagrams form another family of networks in which arcs represent activities. Numbers on the arcs represent the times to perform them, and the nodes are states achieved as the result of the completion of the activities represented by their incoming arcs.

A specimen network is the following where the values marked on the arcs are the maximum flows along them:

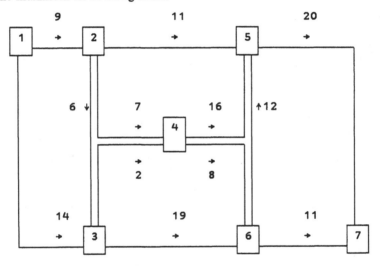

Such a network may be represented by an N by N matrix where N is the number of nodes, and the value in the cell (r,c) is the directed distance between node r and node c. No connection between the nodes r and c is represented by a 0 in the cell (r,c). Thus the matrix representing the above network is

```
     NET
0 9 14 0  0  0  0
0 0  6 7 11  0  0
0 0  0 2  0 19  0
0 0  0 0 16  8  0
0 0  0 0  0  0 20
0 0  0 0 12  0 11
0 0  0 0  0  0  0
```

Arcs are represented by non-zero entries, and it may be assumed without loss of generality that for networks with a single source and a single sink, the former is node 1, and the latter the node represented by the last row/column. Thus the first column and the last row of such a network matrix must consist of all zeros.

Typically network matrices are sparse and an alternative space-saving representation of such matrices is a two-item nested vector, the first item of which is a binary matrix of node connectivities **NETC**, and the second item is a vector **NETV** of the values of the non-zero items in row-major order. The above network could then be represented by the nested vector

```
      NETC NETV
0 1 1 0 0 0 0   9 14 6 7 11 2 19 16 8 20 12 11
0 0 1 1 1 0 0
0 0 0 1 0 1 0
0 0 0 0 1 1 0
0 0 0 0 0 0 1
0 0 0 0 1 0 1
0 0 0 0 0 0 0
```

Conversion to the **NET** form is achieved by **FULNET NETC NETV** where **FULNET** is

```
[0]     Z←FULNET R
[1]     Z←↑R
[2]     ((,Z)/,Z)←2⊃R
```

6.4.1 The Vector of Paths through a network

Consider the problem of determining all paths from the source node to the sink node. Assume that L is a binary connectivity matrix such as **NETC** representing a network without loops, and that the items of R represent node numbers. These paths can be determined by the pair of linked recursive functions **OUTFROM** and **ROOT**:

```
[0]     Z←L OUTFROM R
[1]     →L1 IF 0=ρ,R                    ⍝ branch if list of nodes empty
[2]     →0 Z←(L ROOT↑R),L OUTFROM 1↓R ⍝ process first node and recurse
[3]     L1:Z←R                          ⍝ stopping action

[0]     Z←L ROOT R;T
[1]     →L1 IF~∨/T←L[R;]                ⍝ branch if all-zero row found
[2]     →0 Z←R,¨L OUTFROM T/⍳↑ρL        ⍝ join current node to all lower
[3]     L1:Z←R                          ⍝ stopping action
```

Line [2] of **OUTFROM** says

```
        (L ROOT↑R),     ⍝ find all the paths which proceed downwards from
                          the first item in R, and join them on to ...
        L OUTFROM 1↓R   ⍝ ... ditto for the rest of R
```

ROOT then tests whether the all-zero row (i.e. the sink) has been found, and if not joins the current node R to each of the trees which spread out from the

nodes to which R is connected. OUTFROM thus corresponds to processing "along"
a vector of nodes, ROOT to processing "down" from a single node.

```
      ,[ι0]NETC OUTFROM 1
 1 2 3 4 5 7
 1 2 3 4 6 5 7
 1 2 3 4 6 7
 1 2 3 6 5 7
 1 2 3 6 7
 1 2 4 5 7
 1 2 4 6 5 7
 1 2 4 6 7
 1 2 5 7
 1 3 4 5 7
 1 3 4 6 5 7
 1 3 4 6 7
 1 3 6 5 7
 1 3 6 7
```

The ,[ι0] transforms the vector of paths to a one-column matrix which makes
it easier to read.

If a network has loops, it is necessary to carry into the recursion a list of
nodes encountered so far. In the following version of OUTFROM and ROOT, L is a
two-item vector where the first item is the enclosure of the connectivity matrix,
e.g. NETC, and the second item is the list of nodes encountered so far.

```
[0]    Z←L OUTFROML R
[1]    →L1 IF 0=ρ,R
[2]    →0 Z←((L,↑R)ROOTL↑R),L OUTFROML 1↓R
[3]    L1:Z←R

[0]    Z←L ROOTL R;T
[1]    →L1 IF~∨/T←R⌷[1]↑L
[2]    →0 Z←R,¨L OUTFROML(T/ι↑ρ↑L)~1↓L
[3]    L1:Z←R
```

The **without** in the second line of ROOTL inhibits the processing of nodes which
have already been visited.

For networks without loops, such as NETC, the only difference between
OUTFROM and OUTFROML is that the left argument of the latter must be
enclosed, that is

```
      (⊂NETC)OUTFROML 1
```

is equivalent to

```
      NETC OUTFROM 1
```

An example of a network with loops is

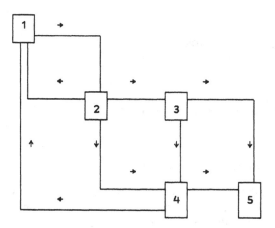

for which the connectivity matrix **NETL** and the result of **OUTFROML** are the following:

```
       NETL
0  1  0  0  0
1  0  1  1  0
0  0  0  1  1
1  0  0  0  1
0  0  0  0  0
```

```
      (⊂NETL)OUTFROML 1
 1 2 3 4 5  1 2 3 5  1 2 4 5
```

6.4.2 Parallel computation along paths

The function **PV** (Path to Value) converts a path **R** into a vector of arc values lying on the path. The argument **L** is the matrix representation of a network.

```
[0]    Z←L PV R
[1]    Z←(2,/R)⌷¨⊂L
```

```
      NET PV 1 2 3
9 6
```

The expression **+/NET PV 1 2 3 4 5 6 7** thus adds arc values along the first path in **NETC OUTFROM 1**.

Another way of looking at this problem is to extend the functions **OUTFROML** and **ROOTL** to obtain **ADDFROM** and **ADDROOT** in which the addition of values along the arcs takes place as the arcs are encountered :

```
[0]    Z+L ADDFROM R
[1]    →L1 IF 0=ρ,R
[2]    →0 Z+(L ADDROOT↑R),⊂L ADDFROM 1↓R
[3]    L1:Z+R


[0]    Z+L ADDROOT R;T
[1]    →L1 IF~∨/T+R[[1]L≠0
[2]    →0 Z+((R,¨T)[]¨⊂L)+(⊂L)ADDFROM¨T+T/ι↑ρL
[3]    L1:Z+0


       εNET ADDFROM 1
53 57 36 66 45 52 56 35 40 52 56 35 65 44
```

Addition could be replaced by other scalar dyadic functions, e.g. the minimum function. An alternative to defining another pair of functions is to convert ADDFROM and ADDROOT to operators, thereby generalizing the function + in the middle of line [2] of ADDROOT. This amendment also requires adjustment to line [3] of ADDROOT to contain the identity item of the function P.

```
[0]    Z+L(P FROM)R
[1]    →L1 IF 0=ρ,R
[2]    →0 Z+(L(P FROOT)↑R),⊂(P FROM)1↓R
[3]    L1:Z+R


[0]    Z+L(P FROOT)R;T
[1]    →L1 IF~∨/T+R[[1]L≠0
[2]    →0 Z+((R,¨T)[]¨⊂L)P(⊂L)(P FROM)¨T+T/ι↑ρL
[3]    L1:Z+P/ι0


       NET
0  9  14  0   0   0   0
0  0   6   7  11   0   0
0  0   0   2   0  19   0
0  0   0   0  16   8   0
0  0   0   0   0   0  20
0  0   0   0  12   0  11
0  0   0   0   0   0   0


       εNET +FROM 1
53 57 36 66 45 52 56 35 40 52 56 35 65 44
       εNET ⌊FROM 1
2 2 2 6 6 7 7 7 9 2 2 2 12 11
```

The minimum of the sums along all paths is the shortest path-length through the network, and the maximum of these sums is the longest path-length, which in the case of a PERT network is that of the critical path. There is an analogy here with inner products, in that one function is performed along all paths, and the reduction of a second function provides a quantity of interest. This leads to the definition of an operator NIP standing for "Network Inner Product."

```
[0]    Z+(P NIP Q)R
[1]    Z+P/εR(Q FROM)1
```

Thus

```
      ⌊NIP+NET
35
```

is the length of the shortest path of **NET**, and

```
      ⌈NIP+NET
66
```

is the length of the longest/critical path of **NET**. Assuming that node 1 is the source, the network may be taken as the single argument of the operator **ROUTE**:

```
[0]   Z←(P ROUTE)R
[1]   Z←((R≠0)OUTFROM 1)[(Z=P/Z)/⍳⍴Z←∊R+FROM 1]

      ⌊ROUTE NET
 1 2 4 6 7  1 3 4 6 7
```

is then the shortest path or paths of **NET**, and

```
      ⌈ROUTE NET
 1 2 3 6 5 7
```

is the longest/critical path or paths of **NET**.

6.4.3 Assignment of Flows

Suppose that the values on the arcs of a network matrix represent non-negative capacities and that as flow is sent down a path from source to sink the capacities on the component arcs in the route are reduced by the amount of the flow. As before it is assumed that the source and sink nodes are the first and last nodes respectively. First **MSUB** (standing for "Matrix Subtract") is constructed which has a network matrix as its left argument and whose right argument is a two-item vector, the first item of which is the row and column indices of an arc and the second an amount to be subtracted from the capacity of that arc. For example **NET** and the matrix resulting from subtracting 2 from **NET[1;2]** are shown below.

	NET							(NET MSUB(1 2)2)					
0	9	14	0	0	0	0		0	7	14	0	0	0	0
0	0	6	7	11	0	0		0	0	6	7	11	0	0
0	0	0	2	0	19	0		0	0	0	2	0	19	0
0	0	0	0	16	8	0		0	0	0	0	16	8	0
0	0	0	0	0	0	20		0	0	0	0	0	0	20
0	0	0	0	12	0	11		0	0	0	0	12	0	11
0	0	0	0	0	0	0		0	0	0	0	0	0	0

Using the technique of **CHANGE** in Section 1.4.2 leads to

```
[0]   Z←L MSUB R
[1]   (Z[L])←((Z←↑R)[L])-2⊃R
[2]   Z←L
```

The function **PSUB** standing for "Path Subtract" extends this to an entire path. R is the path catenated to the amount to be subtracted.

```
[0]    Z←L PSUB R
[1]    →L1 IF 1=ρ↑R
[2]    →0 Z←(L MSUB(⊂2↑↑R),2⊃R)PSUB 1 0↓¨R
[3]    L1:Z←L
```

so to subtract 2 from the path 1 2 3 4 5 7 use as right argument the path/amount vector (1 2 3 4 5 7)2:

```
      NET PSUB(1 2 3 4 5 7)2
0 7 14 0  0   0  0
0 0  4 7 11   0  0
0 0  0 0  0  19  0
0 0  0 0 14   8  0
0 0  0 0  0   0 18
0 0  0 0 12   0 11
0 0  0 0  0   0  0
```

The function FLUX has as its right argument a vector of paths, and progressively subtracts the maximum possible flow along each path in turn. The maximum flow along any path is the minimum capacity of the arcs in that path.

```
[0]    Z←L FLUX R
[1]    →L1 IF 0=ρR
[2]    →0 Z←(L PSUB(⊂↑R),⌊/L PV↑R)FLUX 1↓R
[3]    L1:Z←L
```

```
      NET FLUX NETC OUTFROM 1
0 0 0 0   0 0 0
0 0 0 4  11 0 0
0 0 0 0   0 1 0
0 0 0 0  11 8 0
0 0 0 0   0 0 3
0 0 0 0   0 0 5
0 0 0 0   0 0 0
```

Finally the function ALLOC combines the roles of OUTFROM in detecting paths, and FLUX in sending flow along them and also attempts to maximize the total flow through the network. It is known that the objective of achieving a maximum network flow is best promoted by sending flow as far as possible along paths with small numbers of arcs. Therefore the vector of paths corresponding to a network should thus be ordered from least to greatest numbers of arcs, which is achieved by the function UPG.

```
[0]    Z←ALLOC R
[1]    Z←R FLUX UPG(R≠0)OUTFROM 1

[0]    Z←UPG R
[1]    Z←R[⍋∊ρ¨R]

       ALLOC NET
0 0 0 0  0 0 0
0 0 6 7  2 0 0
0 0 0 0  0 7 0
0 0 0 0 14 8 0
0 0 0 0  0 0 8
0 0 0 0 11 0 0
0 0 0 0  0 0 0
```

6.4.4 Minimum Spanning Tree

An undirected network is one in which the arcs do not have directions associated with them and thus its matrix representation must be symmetric. A minimum spanning tree of such a network is a tree structure which includes all the nodes and for which the sum of total arc values is a minimum. This might for example be of interest to an oil prospecting company anxious to minimize the total length of pipeline needed to connect a given set of oil wells. Such a structure may not *look* like a tree in nature or in the sense of the previous diagrams, however all that is required to give the tree property to an assembly of arcs and nodes is that it be connected, and that the number of arcs is one less than the number of nodes. NET may again be used as an example, but since it now represents a graph with undirected nodes it must be made symmetric by

```
      SYMNET←NET+⍉NET
      SYMNET
 0  9 14  0  0  0  0
 9  0  6  7 11  0  0
14  6  0  2  0 19  0
 0  7  2  0 16  8  0
 0 11  0 16  0 12 20
 0  0 19  8 12  0 11
 0  0  0  0 20 11  0
```

To develop a minimum spanning tree algorithm, a tree is modelled as a vector of two-item vectors representing arcs, each of which consists of a left and a right node. If the network is the left argument L of a function MST which returns the minimum spanning tree as a vector of arcs, then the number of items in the result of MST must be $^{-}1+\uparrow\rho L$ by the definition of a tree.

The principle of the algorithm is that at any intermediate stage of constructing the tree, the nodes fall into two disjoint sets, U which includes the nodes used so far, and V those *not* used so far. The next arc to be added to the tree is the lowest-value arc (LVA) connecting these two sets, or the first such arc if there are several of equal value. The right node of the lowest-value arc is the new node to be added to the set U and removed from the set V.

Assume that a function `LVA` has been defined which produces the lowest-value arc for a network `L` and a node-list `R` corresponding to the set `U`. The revised tree after one step of the process is

```
(⊂T←L LVA R),   ⍝ the new arc (enclosed) joined to...
L MST           ⍝ ... the algorithm applied to ...
R               ⍝ ... the previous node list with ...
,1↓T            ⍝ ... the right node of LVA appended.
```

The process starts with any node and stops when `(⍴R)=↑⍴L`, i.e all the nodes are included in the node list. The complete function is:

```
[0]    Z←L MST R;T
[1]    →L1 IF (⍴R)=↑⍴L
[2]    →0 Z←(⊂T),L MST R,1↓T←L LVA R
[3]    L1:Z←⍳0
```

The problem has been reduced to that of calculating the `LVA` which in turn consists of finding the row and column indices of the first occurrence of the minimum non-zero value (`MNZ`) in the set `L[U;V]`. This is given by

```
↑ONES T=MNZ/,T←L[U;V]
```

where `ONES` is the auxiliary function defined in Section 4.2.1, which returns a vector of co-ordinate pairs corresponding to the position of the 1s in a binary matrix:

```
[0]    Z←ONES R
[1]    Z←⊂[1]1+(⍴R)⊤¯1+(,R)/⍳×/⍴R
```

and `MNZ` returns `L⌊R` unless either is zero in which case it returns the other.

```
[0]    Z←L MNZ R
[1]    →L1 IF ∨/0=L,R           ⍝ branch if either L or R is zero
[2]    →0 Z←L⌊R
[3]    L1:Z←L+R
```

These lead to the definition of `LVA` as

```
[0]    Z←L LVA R;T;U;V
[1]    Z←L[U;V←(⍳↑⍴L)~U←∊R]
[2]    Z←(↑ONES Z=MNZ/,Z)⌷¨U V
```

The minimum spanning tree of `SYMNET` is then

```
     SYMNET MST 1
1 2  2 3  3 4  4 6  2 5  6 7
```

The *value* of `MST`, i.e. the sum of the values of its constituent arcs, is

```
[0]    Z←MSTV R
[1]    Z←+/(R MST 1)⌷¨⊂R
```

```
     MSTV SYMNET
```

47

6.4.5 Precedence and Reachability

A further problem which might arise in modelling the sort of networks which have been the subject of the preceding Sections is finding how far it is possible to travel in a given number of steps.

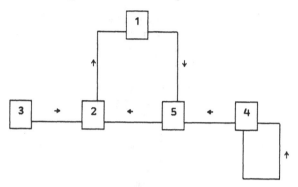

For a graph such as the above which may be cyclic and contain self-looping nodes, and whose connectivity matrix is L, the *precedence matrix* answers the question what nodes may be reached in *exactly* R steps, where R is a non-negative integer. The *reachability matrix* answers the question what nodes may be reached in R steps or less. The precedence matrix is an extension of the connectivity matrix. The two are equivalent when R = 1, and R = 0 denotes the identity matrix. Calculation of the nodes which can be reached at the next step provides in line [2] of the function PREC an application of the ∨.∧ inner product.

```
[0]    Z←L PREC R
[1]    →L1 IF R=0
[2]    →0 Z←(L PREC R-1)∨.∧L
[3]    L1:Z←ID↑ρL
```

```
[0]    Z←ID R;T
[1]    Z←T∘.=T←ιR
```

For one node to be reachable from another in R steps requires either R-step precedence or that reachability has already been achieved in fewer steps, hence line [2] of REACH:

```
[0]    Z←L REACH R
[1]    →L1 IF R=0
[2]    →0 Z←(L PREC R)∨L REACH R-1
[3]    L1:Z←ID↑ρL
```

The matrix CM is the connectivity matrix of the above graph and is used to illustrate the calculation of precedence and reachability matrices.

```
        CM
0 0 0 0 1
1 0 0 0 0
0 1 0 0 0
0 0 0 1 1
0 1 0 0 0
```

```
        CM PREC 1                              CM REACH 1
0 0 0 0 1                               1 0 0 0 1
1 0 0 0 0                               1 1 0 0 0
0 1 0 0 0                               0 1 1 0 0
0 0 0 1 1                               0 0 0 1 1
0 1 0 0 0                               0 1 0 0 1
        CM PREC 2                              CM REACH 2
0 1 0 0 0                               1 1 0 0 1
0 0 0 0 1                               1 1 0 0 1
1 0 0 0 0                               1 1 1 0 0
0 1 0 1 1                               0 1 0 1 1
1 0 0 0 0                               1 1 0 0 1
        CM PREC 3                              CM REACH 3
1 0 0 0 0                               1 1 0 0 1
0 1 0 0 0                               1 1 0 0 1
0 0 0 0 1                               1 1 1 0 1
1 1 0 1 1                               1 1 0 1 1
0 0 0 0 1                               1 1 0 0 1
```

Exercises 6b

1. Write a function MAKENET to construct a connectivity matrix of dimensions given by the left argument for a vector of co-ordinate vectors, e.g.

```
    4 4 MAKENET (1 2)(2 2)(3 1)(3 4)
```

should return

```
0 1 0 0
0 1 0 0
1 0 0 1
0 0 0 0
```

2. For the connectivity matrix NETL how would use APL2 to answer the questions:

    (a) To how many nodes are there routes from nodes 1,2,3 and 4?

    (b) How many nodes can be reached in exactly three steps from node 3?

3. How would you display the minimum spanning tree of SYMNET in Section 6.4.4 with the values of the arcs displayed beneath them thus:

```
  1 2   2 3   3 4   4 6   2 5   6 7
    9     6     2     8    11    11            ?
```

4. The table below gives the distances between each of a set of five towns:

```
0 6 7 9 3
6 0 5 2 4
7 5 0 1 8
9 2 1 0 3
3 4 8 3 0
```

What is the minimum length of a road network which ensures that every town is reachable from every other?

Summary of Operations used in Chapter 6

Trees

Section 6.2
PATH   path to item in tree with keys

Section 6.2.1
ANCIN   ancestors of item in tree with keys

Section 6.2.2
SUBT   subtree determined by given key
STPATH   path to subtree

Section 6.2.3
CUTFROM   removes subtree
SWAP   exchanges subtrees

Section 6.3
INS   inserts in binary tree
MAKET   makes binary tree

Section 6.3.2
ISIN   tests for item in binary tree

Section 6.3.5
GT   generalized "greater than"

Exercises 6a
SUB   finds subtree at given node in binary tree

Networks

Section 6.4
FULNET   network conversion from nested vector to simple matrix

Section 6.4.1
OUTFROM   all paths from a vector of nodes
ROOT   all paths from a given node
OUTFROML   enhancement of OUTFROM to deal with loops
ROOTL   enhancement of ROOT to deal with loops

Section 6.4.2

PV	converts path to vector of values along arcs
ADDFROM	sums along all paths from a vector of nodes
ADDROOT	sums along paths from given node
FROM	operator extension of ADDFROM
FROOT	operator extension of ADDROOT
NIP	"network inner product" applicable to e.g. PERT network
ROUTE	path satisfying e.g. shortest or longest path criterion

Section 6.4.3

MSUB	subtract value from single item in matrix
PSUB	subtract value from each arc in path
FLUX	subtracts values progressively from vector of paths
ALLOC	allocates maximum flow through network
UPG	sorts paths in order of increasing length

Section 6.4.4

MST	minimum spanning tree
ONES	co-ordinate pairs of 1s in binary matrix
MNZ	minimum non-zero value of a pair of scalars
LVA	lowest value arc connecting two sets of nodes
MSTV	value of minimum spanning tree

Section 6.4.5

PREC	precedence matrix for a directed graph
REACH	reachability matrix for a directed graph
ID	identity matrix

Exercises 6b

MAKENET	constructs connectivity matrix from vector of co-ordinates

Appendix A. Solutions to Exercises

Solutions 1a

1. The expressions are DISPLAYed below, in each case followed by the prototype.

a.   'ABC' 17.6

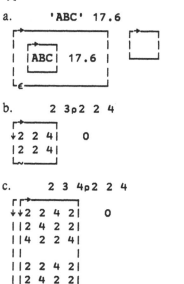

b.   2 3ρ2 2 4

```
┌─────┐
↓2 2 4│   0
│2 2 4│
└─────┘
```

c.   2 3 4ρ2 2 4

```
┌┌──────┐
↓↓2 2 4 2│   0
││2 4 2 2│
││4 2 2 4│
││       │
││2 2 4 2│
││2 4 2 2│
││4 2 2 4│
└└──────┘
```

d.   2 4ρ'ABC' '' ' ' '6'(ι2)(ι0)9 6

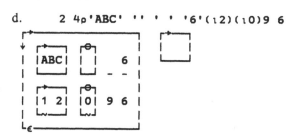

e.   'A' 7.5 5 '5'

```
┌─→─────────┐
│A 7.5 5 5│        _
└─+─────────┘
```

f.   0 3ρ5

```
┌─→─────┐
φ0 0 0│     0
└─∿─────┘
```

g.   0 3ρ'A'

```
┌─→──┐
φ    │        _
└────┘
```

h.   0 3ρ 5 'A'

```
┌─→─────┐
φ0 0 0│     0
└─∿─────┘
```

i.   3 0ρ 5 'A'

```
┌─⊖┐
↓0│    0
│0│
│0│
└─∿┘
```

j.   0 3ρ(5 'A')4

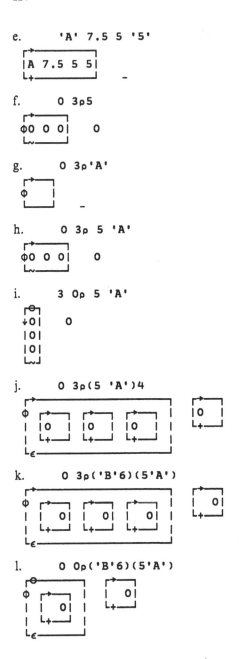

k.   0 3ρ('B'6)(5'A')

l.   0 0ρ('B'6)(5'A')

m.   0 2 0ρ('B'6)(5'A')

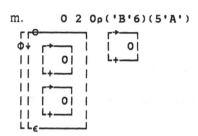

2. 0ρ⊂2ρn and ⊂0 2ρn respectively, where n is any numeric scalar.

3.   [0]   Z←DIS R
      [1]   'SHAPE:'(ρR)'DEPTH:'(≡R)
      [2]   Z←DISPLAY R

4. The value parts of solutions are given in a condensed form of the DISPLAY
format

		shape	depth
a.	(2 3 4)(6 7)	2	2
b.	((4 5)3)('APL')((4 5)3)	3	3

5. (a) is the item-wise product of two two-item vectors; (b) is a three-item vector
in which B×5 is sandwiched between two occurrences of A.

6. (i) d and i   (ii) e and i

7. It gives a DOMAIN ERROR because (ρA)1 is not simple and therefore is not a
valid left argument to ρ.

8. (i) All the same.
  (ii) (c) is in general different, (a) and (b) are the same.
  (iii) (a) and (c) are the same, (b) is different.

9. a. The real part of the product of a complex number and its conjugate is the
square of the magnitude of the argument.

   b. [0]   Z←QUAD R;T
       [1]   Z←-(R[2]+T,-T←((R[2]*2)-×/4,R[1 3])*.5)÷2×R[1]

        ⍉9 11•.○QUAD 1 1 1
‾0.5 ‾0.86603
‾0.5   0.86603

        U←QUAD 1 2J3 4J‾1

Use decode to check the solutions, e.g.

        U[1]⊥1 2J3 4J‾1
8.8818E‾16J‾2.2204E‾16

Solutions 1b

1. (i) No

   (ii) (b),(c),(d) and (f) are the same, namely the three-item depth two vector

   (1 2)(10 20)(3 4).

   (a) is the simple vector 1 2 10 20 3 4, (e) is the three-item depth three vector

   (⊂1 2)(⊂10 20)(⊂3 4).

2. (a) has five items, (b) has four.

3. For conciseness answers are given in a condensed notation in which A is to be read as the array 2 3ρι6.

a. (A) 3   d.   (10×A)30
b. (-A)¯3   e. (A 3) ((2×A)6) ((3×A)9)
c. (A+1)5   f.   (A*2)9

4. 2 3ρ⊂3 5ρ' APL2 IS GREAT'

or 2 3ρ⊂⊃' APL2' ' IS' 'GREAT'

5.   (Z X)←''(ι3)
            DISPLAY Z←Z,⊂X

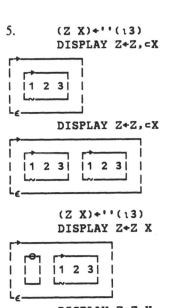

            DISPLAY Z←Z,⊂X

            (Z X)←''(ι3)
            DISPLAY Z←Z X

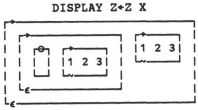

            DISPLAY Z←Z X

None of the four answers are the same.

6. (a) and (c) are the same, namely

(b) and (d) are the same, namely

7. (a) is the simple vector ι3.

(b) is a simple two by three mixed matrix, with a first row of blanks and a second row ι3.

(c) is a simple one-item vector consisting of the character 'X'.

(d) is an empty matrix of shape 2 0. (Z 'X' is a two-item vector and the shape component of its first item is 0, which is brought from the inner to the outer structure.)

8. a. ,'ABC' 'DE' is a two-item vector of character strings,
   ε'ABC' 'DE' is a five-item vector 'ABCDE'.

   b. ,(1 3ρ'ABC')'DE' is a two-item vector, whose first item is a character matrix and whose second is a character string.
   ε(1 3ρ'ABC')'DE' is identical to ε'ABC' 'DE'.

9. a. Calendar for month - weeks horizontal:
```
[0]    Z←SD MONTH DY;HD
[1]    ⍝DY: number of days in month
[2]    ⍝SD: integer indicating start day of month
[3]    HD←'SUN' 'MON' 'TUE' 'WED' 'THU' 'FRI' 'SAT'
[4]    Z←7 7ρHD,(SDρ' '),(ιDY),20ρ' '
```

   b. Calendar for month - weeks vertical:
```
       ⍞3 MONTH 31
SUN      5 12 19 26
MON      6 13 20 27
TUE      7 14 21 28
WED  1   8 15 22 29
THU  2   9 16 23 30
FRI  3 10 17 24 31
SAT  4 11 18 25
```

   c. Calendar for as a character array:

```
      ⌹3 MONTH 31
   SUN MON TUE WED THU FRI SAT
                     1   2   3   4
     5   6   7   8   9  10  11
    12  13  14  15  16  17  18
    19  20  21  22  23  24  25
    26  27  28  29  30  31
```

Notice that formatting results in a decrease in depth:

```
      ≡¨(3 MONTH 31)(⌹3 MONTH 31)
   2 1
```

   d. Vector of start days for each month:

```
[0]    Z←LEAP START_DAY JAN;D
[1]    D←DAYS+12↑0,LEAP
[2]    Z←¯1↓7|+\JAN,D

      DAYS
31 28 31 30 31 30 31 31 30 31 30 31
      0 START_DAY 2
2 5 5 1 3 6 1 4 0 2 5 0
```

   e. Calendar for year shaped by quarters:

```
      SD
2 5 5 1 3 6 1 4 0 2 5 0
      YEAR←SD MONTH¨ DAYS
      ρYEAR                     ⍝ use ⊃YEAR to display as column of months
12
      QTERLY←4 3ρYEAR
```

Solutions 1c

1. In both cases the problem can be approached either by partial enclosure, or by axis-qualified multiplication. For (a) define

```
      A←2 3 4ρ1
      ρ□←⊂[2]A
 1 1 1  1 1 1  1 1 1  1 1 1
 1 1 1  1 1 1  1 1 1  1 1 1
2 4
```

Alternative solutions are

```
      ⊃[2]M×⊂[2]A  and  M×[1 3]A
```

For (b) observe

```
      ρ□←⊂[1 3]A
 1 1 1 1   1 1 1 1   1 1 1 1
 1 1 1 1   1 1 1 1   1 1 1 1
3
```

The alternative solutions are

⊃[1 3]V×⊂[1 3]A, V×[2]A and +\[2]A

2. For economy of space only shape and depth are given in full.

	shape	depth	equivalent to:		shape	depth
a.	ι0	2		j.	(3 4)	2
b.	(3 4)	1		k.	4	2
c.	(3 4)	1	⊃M13	l.	3	2
d.	ι0	2		m.	3	2
e.	4	2		n.	2	2
f.	3	2		o.	2	2
g.	ι0	2	⊂M13	p.	ι0	2
h.	ι0	2	⊂◊M13	q.	ι0	2
i.	(3 4)	1	M13			

Solutions 1d

1. For economy of space solutions values are given in an abridged notation.

	value	shape	depth
a.	E	3	2
b.	2 3ρι6	(2 3)	1
c.	1	ι0	0
d.	(E)	ι0	3
e.	RANK ERROR		
f.	2 3ρι6	(2 3)	1
g.	RANK ERROR		
h.	RANK ERROR		
i.	2	ι0	0
j.	E	3	2
k.	(2 3ρι6)	ι0	2
l.	(2 3ρι6)3	2	2
m.	RANK ERROR		
n.	same as k		
o.	same as k		

2. a. 3 3 3   b. 1   c. 3 3 3 3   d. 3 3
   e. 3   f. 3 1   g. 3   h. 3 3

3.   a. 1 and 3
    b. 1 and 3 have shape ι0; 2 has shape 1 1

4.   a. 2 and 4

    b. In 1, 3, 5 and 7 the shape of the index does not equal the rank of M11. In the case of 6 (1 2) does not match the rank of the second axis of M11.

5. a.   `2⌽[1]M`
 `5 6 7 8`   shape = 4
b.   `2⌽[2]M`
 `2 6 10`   shape = 3
c.   `(⊂2 1)⌽[1]M`
 `5 6 7 8`
 `1 2 3 4`   shape = 2 4
d.   `(⊂2 1)⌽[2]M`
 `2 1`
 `6 5`
`10 9`   shape = 3 2

6. a. `2⌽[1]A`   b. `3⌽[3]A`   c. `2 3⌽[1 3]A`   d. `2 4 3⌽A`

7. a.   `(1+⌿\T=' ')⊂T←'SPARE ME A DIME'`
 `SPARE   ME   A   DIME`

   b.   `(T≠' ')⊂T←,M,' '`

   c.   `∈+\¨(1+0≠V)⊂V`

8. Cardinal to Ordinal

```
[0]   Z←ORDINAL N;T;I
[1]   ⍝N:  simple positive integer
[2]   Z←⍕N
[3]   T←'st' 'nd' 'rd' 'th'
[4]   I←'123'⍳¯1↑Z
[5]   →('1'≠↑¯2↑Z)/OK
[6]   I←4
[7]   OK:
[8]   Z←Z,I⊃T
```

Solutions 1e

1. (a) Yes, value is 3. (b) No, depth is 2.

2. a. `1↓¯1↓(~' '≤V)/V←' ',V,' '`

   b. `⊃(⊂[2]M)~⊂(↑⌽⍴M)⍴' '`

   (Note: If M is non-nested (b) is equivalent to `(∨/M≠' ')/M`.)

3. The results of applying the three collating sequences are:

```
      (M17[CS1▲M17;])(M17[CS2▲M17;])(M17[CS3▲M17;])
ACE   ACE   ace
BAD   ace   bad
BED   BAD   ACE
Bed   BED   bed
CAB   BeD   BAD
DAD   bad   Bed
ace   bed   BED
bad   CAB   dad
bed   DAD   CAB
dad   dad   DAD
```

4. a. (i) ⊃(⊂[2]M)[□AV▲M]   (ii) M[DCS▲M;]

    b. (i)   ⊃Z IF(Z↓Z)=↓ρZ←(⊂[2]M)[□AV▲M]
       (ii)   M[(1,∨/⁻2≠≠M[I;])/I←DCS▲M;]

5. A matrix of 1s of shape ρC.

6. a.   ((1+R-L)ρ1)≤Rρ1

    b.   (1+-C)⌽(1+-W)⊖((1+R-L)ρ1)≤Rρ1

7.   (2 2 2ρ1)≤A13

8. a.   [0]   Z←REPL R
        [1]   Z←R
        [2]   ((' '=∈Z)/∈Z)←'*'

        REPL 2 4ρ'ABC '
ABC*
*ABC

   b. In line 2 replace ' '=Z with 0=Z and '*' with ⊂'N/A'

        Repl 2 4ρ1 2 3 0 0
    1 2 3 N/A
  N/A 1 2   3

Solutions 2a

1.a.   DISPLAY ρV23
┌─┐
│2│
└~┘

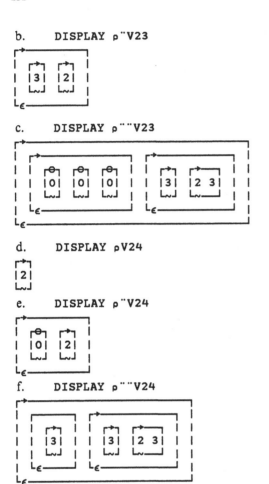

b.   DISPLAY ρ¨V23

c.   DISPLAY ρ¨¨V23

d.   DISPLAY ρV24

e.   DISPLAY ρ¨V24

f.   DISPLAY ρ¨¨V24

2. a.   (24 25 26)   d.   12 6 7 8
   b.   (9 11 13)   e.   12 21
   c.   3 4 5 6 7 8

3.   (↑¨V)←'DA', or in general use ⎕AF.

4. An expression for the weighted moving average is:

    +/¨(⊂W)×(ρW),/V

5. 1-f. 2-b. 3-d. 4-a. 5-g. 6-c.

Solutions 2b

1. a.   `2 3ρ¨W`
   ```
   AB DEF
   ```

   b.   `2 3ρ¨⊂W`
   ```
   ABC DEFG      ABC DEFG ABC
   ```

   c.   `2 3ρ⊂¨W`
   ```
   ABC     DEFG    ABC
   DEFG    ABC     DEFG
   ```

   d.   `(⊂2 3)ρ¨W`
   ```
   ABC   DEF
   ABC   GDE
   ```

   e.   `(⊂2 3)ρ¨⊂W`
   ```
   ABC   DEFG ABC
   DEFG  ABC  DEFG
   ```

   f.   `2 3ρ¨¨W`
   ```
   AA BB CC    DDD EEE FFF GGG
   ```

2. a. `(2 4 1 3)(1 3 2)(1 2 3)`   b. `2 3 1`   c. `3 1 2`

3.   `(?¨3ρ⊂100ρ6)ι¨6`

4. a. Suggested comments are:

```
[0]    Z←L PRT3D R;PLA;ROW;COL
[1]    Z←⊂[2 3]A                 ⍝ make data into a vector of planes
[2]    Z←' ',[1]¨' ',[2]¨Z       ⍝ prefix rows and columns with blanks
[3]    PLA←L[1],¨2⊃L             ⍝ construct plane titles
[4]    ROW←'\',L[3],4⊃L          ⍝ construct row titles
[5]    COL←L[5],6⊃L              ⍝ construct column titles
[6]    Z←(⊂ROW),¨(⊂COL),[1]¨Z    ⍝ attach row & col titles to each plane
[7]    Z←PLA,[1.5]Z              ⍝ attach plane titles
[8]    Z←,[ι0]Z                  ⍝ arrange as a single column
```

   b. The shape/depth table for Z at the various stages of execution is:

	≡	ρ	ρ¨
[1]	2	2	(3 4)(3 4)
[2]	2	2	(4 5)(4 5)
[6]	3	2	(5 6)(5 6)
[7]	3	2 2	3(5 6)
			3(5 6)
[8]	3	2 2 1	3
			(5 6)
			3
			(5 6)

c. The descriptor L[1] is joined to *each* heading and so retains its identity as a single unit. It should therefore remain *enclosed* and so a *cross-section* (indexing) is required. The headings, however, are joined individually, hence *pick* is appropriate.

d. (i) The following changes should be made to TITLES:

```
TITLES[2]←⊂'PLA1' 'PLA2'
TITLES[4]←⊂'ROW1' 'ROW2' 'ROW3'
TITLES[6]←⊂'COL1' 'COL2' 'COL3' 'COL4'
```

(ii) No changes need be made to PRT3D.

5. Pascal triangle

```
[0]    Z←PASCAL N
[1]    Z←1,1,¨(ι¨N)!ιN
[2]    Z←(ι1+N)↑¨Z        ⍝ delete trailing zeros
[3]    Z←⊃⍉¨Z
```

```
[0]    Z←CENTER A;T
[1]    ⍝A: simple character matrix with trailing blanks
[2]    T←+/∧\' '=⌽A
[3]    Z←(-⌊.5×T)⌽A
```

Solutions 2c

1. One possible modification is as follows:

```
[0]    Z←L Compress R;BV;I            .
[1]    (BV I)←L
[2]    Z←BV/[I]R
```

2. Values are:

a.   2-/10 5 2 12 6
5 3 ‾10 6

b.   ‾2-/10 5 2 12 6
‾5 ‾3 10 ‾6

c.   2ρ/2 3 4
 3 3 4 4 4

d.   ‾2ρ/2 3 4
 2 2 2 3 3 3

3 a. Each of the items of V is replaced by the prototype of V which is (0   ) for the test case.

b. ((2×ρ¨V)ρ¨⊂0 ‾1)COMPRESS¨V

4. There are no others. This is discussed in more detail in Section 5.5.3.1.

5. Scan

6. a.   ```
          [0]   Z←DTB R
          [1]   →0 IF(0=ρR)∨' '≠↑⌽Z←R
          [2]   Z←DTB ¯1↓R
          ```

    b. One possibility is to use scan, e.g.

    ```
    [0]   Z←Dtb R
    [1]   Z←⌽(∨\⌽R≠' ')/⌽R
    ```

    c. Use DTB¨VW or Dtb¨VW.

7. a. ×/+/¨2,/ι5

    b. 3 1 4/'ABC'

8.  ```
    [0]   Z←L FIND R;T
    [1]   T←¯1+ρL
    [2]   Z←((¯1+Rι1)↑0),ε(⊂Tρ1),¨(-T)↓¨(+\R)⊂R←L∈R
    ```

9.   ∨/((⊂2 2)ρ¨(1 0 0 1)(1 1 0 1)(1 0 1 1)(1 1 1 1))≤¨⊂MAT

or more briefly

      ∨/((⊂2 2)ρ¨(⊂4 2)⊤¨9 13 11 15)≤¨⊂MAT

Solutions 3a

1. a.   (+/⌿⊃PRICES×STOCKS>0)÷+/⌿⊃STOCKS>0
55.633 24.75 43.3 24 76.5
   b.   (⊃+/⌿⊃PRICES×STOCKS)÷+/⌿⊃STOCKS>0
56.222 24.75 39.108 24.75 76.5

2.   ⍕¨NETMU
 1 2 3   5 1 2 4 3   3 1 4 2

which can be translated into component names by

```
     ⊃(⍋¨NETMU)⌷¨¨⊂⊂CNOS
  X801    X802    X803
  X805    X801    X802    X804    X803
  X803    X801    X804    X802
```

3. The Cash Register System

```
[0]    Z←I RECEIPT STOCK;T
[1]    ⍝STOCK: vector of vectors - each item consists of
[2]    ⍝      (inventory no.)(item name)(unit amount)(costs/unit)
[3]    ⍝I:   vector of inventory numbers in stock
[4]    T←∊↑¨STOCK
[5]    Z←⊃1↓¨STOCK[T⍳I]
```

```
      ⊃STOCK
211 THREADIES   1000 1.98
312 FLATONES       1 1.09
654 LOTSAVOLTS     2 1.55
```

```
      211 654 RECEIPT STOCK
THREADIES   1000 1.98
LOTSAVOLTS     2 1.55
```

Solutions 3b

1. Multiplier is between

```
      ↑⌽((,MTAB)[T])IF .5>+\(,PTAB)[T+⍋,MTAB]
45.993
```

and

```
      ↑((,MTAB)[T])IF .5<+\(,PTAB)[T+⍋,MTAB]
48.173
```

```
2.    +/¨10 12∘.NPV REV REV1 REV2
19005 12510 13697
17947 11896 13002
```

3. The expressions are evaluated for the particular value BANK given in the exercise.

```
a.        ⍴BANK
       4
```

```
b.(i)      +/¨+/¨¨BANK
      15 34 ¯6 25
   (ii)     +/¨+/¨¨0⌈BANK
      34 44 11 25
   (iii)    |+/¨+/¨¨0⌊BANK
      19 10 17 0
```

```
c.          +\¨+/¨¨BANK
      20 15  15 34  ¯6 ¯6  25 25
```

```
d.          +/+/¨¨BANK
      54 14
```

```
e.          +/∊BANK
      68
```

4. Last Trades   a. Find last trade of each stock:

```
[0]     Z←LAST_TRADE X;T;SYM
[1]     T←⍊X
[2]     SYM←T[;1]
[3]     Z←⍊((SYMιSYM)=ιρSYM)/[1]T
```

```
        STP
MMM    3:25  95
T      3:27  36.5
GM     3:31  43
MMM    3:33  42.75
IBM    3:45  102.25
IBM    3:57  102.125
GM     4:02  43.125
GM     4:04  43.375
IBM    4:04  102.25
T      4:05  36.75
IBM    4:12  102.5
```

```
        LAST_TRADE STP
MMM    3:33  42.75
GM     4:04  43.375
T      4:05  36.75
IBM    4:12  102.5
```

b. Last trade of a given stock.

```
[0]     Z←S STK_LAST_TRADE X;T;SYM
[1]     ⍝S: stock symbol character vector
[2]     T←⍊X
[3]     SYM←T[;1]
[4]     Z←((SYMιSYM)=ιρSYM)/[1]T
[5]     T←Z[;1]
[6]     Z←Z[(T~'' '')ιⲤS;]
```

```
        'GM' STK_LAST_TRADE STP
 GM 4:04 43.375
```

c. Enhanced solution if the stock is not traded:

```
[0]     Z←S Stk_last_trade X;T;SYM
[1]     SYM←1⌷[2]T←⍊X
[2]     Z←⍊((SYMιSYM)=ιρSYM)/[1]T
[3]     I←(Z[;1]~'' '')ιⲤS
[4]     →L1 IF I>↑ρZ
[5]     →0 Z←Z[I;]
[6]     L1:Z←'Stock ',S,' not traded'
```

```
        'GM' Stk_last_trade STP
GM 4:04 43.375
        'AA' Stk_last_trade STP
Stock  AA not traded
```

d. Return the last trade after a given time:

```
[0]    Z+TIME TIM_LAST_TRADE X;T;SYM;TIM
[1]    TIME+⍎':.' CHANGE TIME
[2]    TIM+2⎕[2]T+⍎X
[3]    TIM+⍎"(⊂':.')CHANGE"TIM
[4]    SYM+1⎕[2]T+(TIME<TIM)/[1]T
[5]    Z+⊖((SYMιSYM)=ιρSYM)/[1]T

       '4:00' TIM_LAST_TRADE STP
  GM   4:04   43.375
  T    4:05   36.75
  IBM  4:12   102.5
```

(See Section 1.4.2. for **CHANGE**.)

Solutions and Notes 4a

1. In (a) and (b) the right argument is a scalar, and scalar expansion takes place. (b) means apply reshape twice, once with left argument 2, and then with left argument 3.

a.   DISPLAY 2 3ρ⊂V

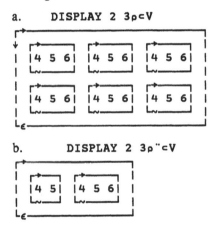

b.   DISPLAY 2 3ρ"⊂V

(c) fails because the numbers of items on each side of the ρ" are not equal.

```
       V+4 5 6
       2 3ρ"V
LENGTH ERROR
       2 3ρ"V
       ∧   ∧
```

In (d) the items of V are scalars and so enclosing each of them makes no difference, that is the **enclose** and **each** cancel each other out and this phrase is exactly the same as 2 3ρV.

d.   `DISPLAY 2 3ρ⊂¨V`

```
┌→─────┐
↓4 5 6│
│4 5 6│
└~─────┘
```

(e) and (f) both fail because the left argument of ρ must be *simple*.

e.   `(⊂2 3)ρV`   f.   `(⊂2 3)ρ⊂V`
`DOMAIN ERROR`   `DOMAIN ERROR`
   `(⊂2 3)ρV`   `(⊂2 3)ρ⊂V`
      `∧      ∧`   `∧      ∧`

In (g) the derived function ρ¨ has a scalar left argument and a vector right argument, so the former is scalar-expanded and item by item execution of ρ results in a three-item vector.

g.   `DISPLAY(⊂2 3)ρ¨V`

```
┌→────────────────────────────┐
│ ┌→─────┐ ┌→─────┐ ┌→─────┐  │
│ ↓4 4 4│ ↓5 5 5│ ↓6 6 6│  │
│ │4 4 4│ │5 5 5│ │6 6 6│  │
│ └~─────┘ └~─────┘ └~─────┘  │
└∊────────────────────────────┘
```

In (h) **reshape-each** has 2 scalar arguments. By the **each** rule both are disclosed prior to function application and the result is enclosed to give finally a depth two result.

h.   `DISPLAY(⊂2 3)ρ¨⊂V`

```
┌→──────────┐
│ ┌→─────┐  │
│ ↓4 5 6│  │
│ │4 5 6│  │
│ └~─────┘  │
└∊──────────┘
```

2 a.   **AB CDE**   e.   **AB CDE**
   b.   **ACDE BCDE**   f.   **ABCDE**
   c.   **AC BD**   g.   **LENGTH ERROR**
   d.   **ABCDE**   h.   **AC AD AE   BC BD BE**

3.   **V**
   A9 B12 B9 b9 B10
   (i)   `⊃(⎕AV⍋⊃V)⌷¨⊂V`
   b9 A9 B12 B10 B9
   (ii)   `V[⎕CS⍋⊃V]`
   A9 B9 b9 B10 B12

4.   `∊(⎕AF(¯1+⎕AF'A')+⍳26),¨⎕AF(¯1+⎕AF'a')+⍳26`

   or   `⎕AF,(0,⍳25)∘.+⎕AF'Aa'`

5.   `∨/∊V1⊆¨⊂V2`   or   `∨/∊V1∊¨⊂V2`

6. a.   `(ιρV)IF∈∧/('AB'≤V)('C'∈(ρV)↑3↓V`

   b.   `Z IF 'C'∈¨(Z←(ιρV)IF'AB'≤V)↓¨⊂V`

7 a. Total words

```
     WORDS←(1↓GETTYSBURG≠' ')⊂GETTYSBURG
     ρWORDS
267
```

b. Number of distinct words

```
     UWORDS←WORDS~¨⊂' ,.;:-'      ⍝ remove punctuation
     ρUWORDS
267
```

```
     ⍝ map upper case to lower case

     FIRST←↑¨UWORDS
     ALPHA←'abcdefghijklmnopqrstuvwxyz'
     ALPHA←ALPHA,'ABCDEFGHIJKLMNOPQRSTUVWXYZ'

     5↑UWORDS
 Fourscore and seven years ago
     (↑¨UWORDS)← ALPHA[26|ALPHAιFIRST]
     5↑UWORDS
 fourscore and seven years ago
```

```
     ⍝ determine number of distinct words

     DISTINCT←((UWORDSιUWORDS)=ιρUWORDS)/UWORDS
     ρDISTINCT
139
```

c. Concordance

```
     ⍝ determine occurrences of each distinct word

     TOTALS←+/DISTINCT∘.≡UWORDS
     ρTOTALS
139
     DISTINCT_CT←DISTINCT,[1.1]TOTALS

     ρDISTINCT_CT
139 2
     SORTED_CT←DISTINCT_CT[▼TOTALS;]
```

```
          10↑[1]SORTED_CT
    that      13
    the       11
    we        10
    to         8
    here       8
    a          7
    and        6
    nation     5
    of         5
    have       5
```

8. It transforms it into a sentence.

Solutions 4b

1.   [0]   Z←DTB R
        [1]   Z←Φ(∨\ΦR≠' ')/ΦR

  a.   [0]   Z←DTBM R
        [1]   Z←DTB¨⊂[2]R

  b.   [0]   Z←L INDEX R
        [1]   Z←((⊂DTB R)≡¨DTBM L)/ι↑ρL

or   [1]   Z←(L∧.=(‾1↑ρL)↑R)/ι¨ρρL

2. a. ⊃≢¨⊂[2]M

  b. ⊃≢¨⊂[⌈/ιρρA]A←',',A

Solutions 4c

1.   A←2 3ρι6
        B←3
        C←'APL'
        E←(⊂A),B,⊂C

a.   DISPLAY 0ρE

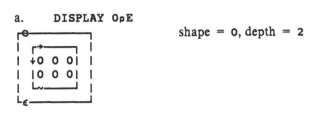

shape = 0, depth = 2

b.   DISPLAY ↑0ρE

shape = 2 3, depth = 1

c.   DISPLAY ↑0ρ⊂E

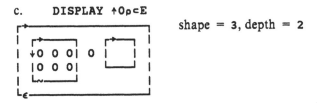

shape = 3, depth = 2

The remaining answers are given in a condensed notation.

	value	shape	depth		value	shape	depth
d.	' '	3	1	g.	3	ι0	0
e.	RANK ERROR			h.	0	ι0	0
f.	3	ι0	0	i.	ι0	0	1

2. (a) and (c) reduce depth, the others do not.
   (c) is subject to INDEX ERROR, (a) is not.
   (b) requires that the rank of A is one or zero.
   (d) and (e) are fully equivalent.

   (i) (a) and (c) are the same, (b),(d) and (e) are the same.
   (Hint - confirm by T•.≡T←(↑A)(1↑A)(1⊃A)(A[1])(1⎕A))

   (ii) all of them are scalar 1 except (b) which is vector 1.

3. The answers are given in a pseudo-APL notation in order to highlight distinctions between e.g. ι0 and the character 'blank'.

a. (i) (0 0)(0)('   ') (ii) (0 0)

b. One is two zeros, the other is three blanks.

c. (i)   (1 2) (3) ('ABC') (0 0) (0 0)
   (ii)   (1 2 0 0 0) (3 0 0 0 0) ('ABC   ')
   (iii)   ('ABC') (3) (1 2) ('   ') ('   ')

d. (i)   1 2 0 (ii) D   (iii)   1 3 A
           3 0 0   2 0 B
           A B C   0 0 C

4. a.   ⊃'THIS' 'IS' 100
             T H I S
             I S
         100 0 0 0

   b.   ⊃⍳¨⍳3

   c.   ⊂[2 1]⊃V47

5. ‾1↑ always returns a one item vector *containing* the first item; ↑⌽ returns the last item itself.

6. a. 1   b. 1 2 3   c. 1 4   d. 1 2 3

7.   ⌽⍴M or ⍴⌽M in both cases.

8.   ↑¨V48[⍋2⊃¨V48] or ↑¨V48[⍋2⊃¨V48]
   GETTY TRUMP

9. first

Solutions 4d

1 a.   DISPLAY ×/2 3 0⍴0
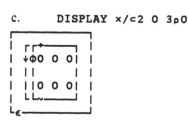

   b.   DISPLAY ×/2 0 3⍴0

   c.   DISPLAY ×/⊂2 0 3⍴0

2 a.   DISPLAY ↑(⍳0)↓0⍴⊂2 3⍴0

```
┌→─────┐
↓0 0 0│
│0 0 0│
└~─────┘
```

b.   DISPLAY ↑⊞¨0⍴⊂2 3⍴0

```
┌→───┐
↓0 0│
│0 0│
│0 0│
└~───┘
```

c.   DISPLAY ↑(⊂3 4)⍴¨0⍴⊂2 3⍴0

```
┌→───────┐
↓0 0 0 0│
│0 0 0 0│
│0 0 0 0│
└~───────┘
```

Some implementations may give the same answers to parts (b) and (c) as to part (a).

3 a.   DISPLAY ⍴/2 3 0⍴0

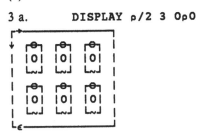

b.   DISPLAY ⍴/0⍴⊂2 3⍴0

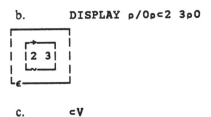

c.   ⊂V

4 a.   DISPLAY ⍉/0⍴⊂2 3⍴0

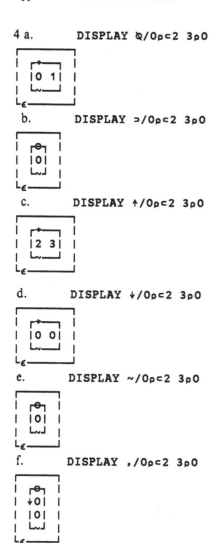

b.   DISPLAY ⊃/0⍴⊂2 3⍴0

c.   DISPLAY ↑/0⍴⊂2 3⍴0

d.   DISPLAY ↓/0⍴⊂2 3⍴0

e.   DISPLAY ~/0⍴⊂2 3⍴0

f.   DISPLAY ,/0⍴⊂2 3⍴0

5 a.   DISPLAY ρ↑,/Oρ⊂2 9ρ0

```
┌→──┐
│2 0│
└~──┘
```

   b.   DISPLAY ρ↑,/Oρ⊂2 9 9ρ0
               DISPLAY ρ↑,/Oρ⊂2 9 9ρ0

```
┌→────┐
│2 9 0│
└~────┘
```

   c.   DISPLAY ρ↑,/Oρ⊂2 9 9 9ρ0

```
┌→──────┐
│2 9 9 0│
└~──────┘
```

Eventually what happens is

```
          DISPLAY ρ↑,/Oρ⊂2 9 9 9 9 9 9 9ρ0
WS FULL
          DISPLAYρ↑,/Oρ⊂2 9 9 9 9 9 9 9ρ0
                  ∧                   ∧
```

Solutions 5a

1. One way to write the operator CONSEC is

```
[0]    Z←L(P CONSEC)R
[1]    Z←Z/ιρZ←((L-1)ρ1)⊆2 P/R

       V
2 3 4 3 4 5 2 2 7
       2<CONSEC V      ⍝for strictly increasing sequences
1 2 4 5 8
       3≥CONSEC V      ⍝for non-decreasing sequences, etc.
6
```

2. a. It is convenient to write the operator BASE on the assumption that two functions TODEC and FROMDEC exist to convert to and from decimal notation.

```
[0]    Z←L(P BASE Q)R
[1]    (L R)←Q TODEC¨L R
[2]    Z←((ρZ)ρ10)⊥Z←Q FROMDEC L P R
```

The functions TODEC and FROMDEC can then be written:

```
[0]     Z←L FROMDEC R
[1]     Z←((1+⌈L⍟R)⍴L)⊤R

[0]     Z←L TODEC R
[1]     Z←((⍴Z)⍴L)⊥Z←10 FROMDEC R

        16+BASE 7 23
42
        1111÷BASE 2 11
101
```

b. Extend to process arrays by using each, e.g.

```
        □←A←2 2⍴1111 110 10010 100001
 1111       110
10010 100001

        A÷BASE 2¨11
101    10
110 1011
```

3. **ROOT** is a dyadic function. **ROOTOP** achieves the identical result for scalar arguments by producing at an intermediate stage a monadic derived function which could be called the Pth root-function. Using **ROOTOP** can sometimes avoid the need for enclosure, for example:

```
        2 3 ROOTOP¨⍳5
 1 1   1.4142 1.2599   1.7321 1.4422   2 1.5874   2.2361 1.71
        2 3 ROOT¨⍳5
LENGTH ERROR
        2 3 ROOT¨⍳5
        ∧         ∧
        (⊂2 3)ROOT ⍳5
 1 1   1.4142 1.2599   1.7321 1.4422   2 1.5874   2.2361 1.71
```

4. b and c

Solutions 5b

1. a. **PRODUCT** : Apart from the function name the only change required is to replace + by ×.

b. **JOIN** : The changes required here are more subtle and require the use of ⊂ and ↑ on account of the non-pervasiveness of **catenate**.

```
[0]     Z←JOIN R
[1]     →L1 IF 1=⍴R
[2]     →0 Z←(↑R),⊃JOIN 1↓R
[3]     L1:Z←⊂↑R
```

2. Change line 4 of **Path** so that the function reads:

```
[0]    Z+L PAth R;T
[1]    →L1 IF∧/(L∊∊R),1≤≡R
[2]    →0 Z+⍳0
[3]    L1:→L2 IF(⍳0)≡ρR
[4]    T+⊂1+(ρR)T¯1+(,L∊¨∊¨R)⍳1
[5]    →0 Z+T,L PAth T⊃R
[6]    L2:Z+(⊂⍳0),L PAth↑R
```

3. A suitable function to change all occurrences is

```
[0]    Z+L CHALL R
[1]    →L1 IF~(↑L)∊∊R
[2]    →0 Z+L CHALL L CHANGE R
[3]    L1:Z+R
```

4. a. The following is a recursive definition of the **POWER1** operator:

```
[0]    Z+L(P POWER1 Q)R
[1]    →L1 IF Q=0
[2]    →0 Z+(L P POWER1(Q-1)R)P R
[3]    L1:Z+L
```

```
       3 *POWER1 3 2
6561
       12×POWER1 3 2
96
```

b. The following is a recursive definition of the **POWER2** operator:

```
[0]    Z+L(P POWER2 Q)R
[1]    →L1 IF Q=0
[2]    →0 Z+L P L P POWER2(Q-1)R
[3]    L1:Z+R
```

c. The following sequence illustrates the convergence of the iterative solution of the equation y = cos(y):

```
       2○POWER2 10 1
0.7442373549
       2○POWER2 25 1
0.7390713653
       2○POWER2 50 1
0.7390851339
       2○POWER2 100 1
0.7390851332
```

d. Use **POWER2** as follows:

```
    CRYPT+CODE CODIFY POWER2 4 'DOG'
```

The receiver decodes using

```
    CODE DECODE POWER2 4 CRYPT
DOG
```

where **DECODE** is given by

```
[0]    Z←L DECODE R
[1]    Z←ALF[L⍳R]
```

5. An operator POLISH and compatible functions MEAN and MEDIAN are:

```
[0]    Z←(P POLISH)R
[1]    Z←-(∈P¨⊂[1]R)-[2]R←R-[1]∈P¨⊂[2]R
```

```
[0]    Z←MEAN R
[1]    Z←(+/R)÷⍴R
```

```
[0]    Z←MEDIAN R
[1]    Z←.5×+/R[⌈.5×0 1+⍴R←R[⍋R]]
```

```
       T
0  6  6
4  0  2
       MEAN POLISH T
¯3   2   1
 3  ¯2  ¯1
       MEDIAN POLISH T
¯4   1  0
 4  ¯1  0
```

Solutions and Notes 5c

```
1.      ,/H
 1   8   9 10    6  5  4  3  9  9
 2  11  12 13    2  1  6  5  9  9
      ⊂[2]H
 1    8   9 10    6  5  4  3    9  9
 2   11  12 13    2  1  6  5    9  9
```

```
2. a.           ,/M
      ABC DEF
   b.           ,/¨M
      ABC
      DEF
```

```
   c.           ,/A
      ABC DEF
      GHI JKL
   d.           ,/¨A
      ABC
      DEF

      GHI
      JKL
```

```
3. a.  [0]    Z←L SUBMAT R
       [1]    Z←(⊂L)⍴¨L[1],/[1]L[2],/R
```

   b. Define

```
[0]    Z←TEST R
[1]    Z←∧/(~2 2⍴R),(1 2)(2 3)(3 2)⎕¨⊂R
```

The given pattern can therefore be tested for by

```
      TEST¨3 3 SUBMAT M54
0 1 0
0 0 1
```

c. Generalizing this test to cover *any* pattern requires the definition of an APL2 object to represent a pattern. One possibility is to use a three item vector, the first item of which is the **shape** of the pattern, the second item is a vector of the co-ordinates of the 0s, and the third item a vector of the co-ordinates of the 1s. The pattern above would then be represented by

```
      PAT←(3 3)(⊂2 2)((1 2)(2 3)(3 2))
```

Care has to be taken to ensure that the two vectors of coordinate vectors are of equal depth, hence the explicit **enclose** in the second item. A function which tests for the occurrences of a binary pattern L in a binary matrix R is

```
[0]    Z←L PATIN R
[1]    R←(↑L)SUBMAT R
[2]    Z←⊃∧/∧/¨¨(~L[2]SEL¨R)(L[3]SEL¨R)
```

```
[0]    Z←L SEL R
[1]    Z←L⎕¨⊂R
```

Hence the test above could be achieved by

```
      PAT PATIN M54
0 1 0
0 0 1
```

4. a.   (≠\(3×⍴CV)⍴1 1 0)\CV

   b.   (≠\(1.5×⍴CV)⍴1 0 1)\CV

   c.   (∧\'∩'=LINE)/LINE

5. a. First observe that the shape vector rule requires that the outer structure be a two by two matrix. To determine the items, e.g. first row, second column, the **each** rule must be applied, i.e. the vector 2 2 and the matrix 4 1⍴'ABCD' are both disclosed, ⍴ is applied, and the result enclosed as a scalar to take its place as item [1;2] of the result. Since none of the items so enclosed exceeds depth one, the overall depth of the outer product is two. The final result is therefore

```
      ((2 2)3)∘.⍴6(4 1⍴'ABCD')
 6 6      AB
 6 6      CD

 6 6 6    ABC
```

b. Start by displaying the two matrices A and B:

```
      DISPLAY A B
```

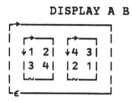

The APL2 rule says enclose along last and first axes, and take the outer product:

```
      DISPLAY¨(⊂[2]A)(⊂[1]B)
```

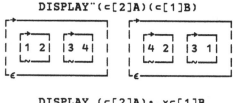

```
      DISPLAY (⊂[2]A)∘.×⊂[1]B
```

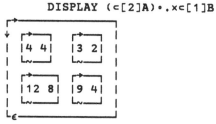

The numbers appearing in the above display are clearly those which arise in "row-into-column" evaluation of the matrix product. The result is completed by taking the sums of the numbers within each inner box:

```
      DISPLAY +/¨(⊂[2]A)∘.×⊂[1]B
┌→────────┐
↓  8    5|
| 20  13|
└~────────┘
```

which is the same as

```
      DISPLAY A+.×B
┌→────────┐
↓  8    5|
| 20  13|
└~────────┘
```

```
6. a.       2 4•.+1 4 6
      3 6  8
      5 8 10
   b.       2 4•.,1 4 6
      2 1  2 4  2 6
      4 1  4 4  4 6
   c.       2 2•.⍴⍳4
      1 1  2 2  3 3  4 4
      1 1  2 2  3 3  4 4
   d.       2 3•.⍴⍳4
      1 1    2 2    3 3    4 4
      1 1 1  2 2 2  3 3 3  4 4 4
   e.       2 4 6•.,'AB'
      2 A  2 B
      4 A  4 B
      6 A  6 B
   f.       2 4 6•.,'AB' 'CDE'
      2 AB  2 CDE
      4 AB  4 CDE
      6 AB  6 CDE

7. a.    3 2 1⍴.⍴3 2 1
    3 2
    1 3
    2 1

   b.    3 2 1⍴.⍴3 2 1
      2 2 2  3
      1 1    2 2 2
      3      1 1

   c.    1 2 3,.⍴4 5 6
      4 5 6  4 5 6
   d.    1 2 3⍴.,4 5 6
      6
   e.    1 2 3~.+2 3 4
      3
   f.    1 2 3+.~2 3 4
      1
```

8. a. 1.875 in both cases.   Result is weighted average with weights in
                                     descending powers of 2, viz. 4 2 1 1.

   b.   2   2.5 3   in both cases - table of averages by pairs
        2.5 3   3.5

   c.   3 3.5 4   for AVG - ½ of 1 2 3,1 2 4, and 1 2 5.

        3 4   3.5 4.5   4 5   for MID - (⊂1 2) added to ½ of
                                     -/3 1 2, -/4 1 2, and -/5 1 2.

d.   **7.5**   for **AVG** - **.5×+/ι5**

     3.5   4.5   for **MID** - (**⊂1 2**) added to ½ of **-/3 4 5 1 2**

e.   **1.9375** in both cases.   Result of intermediate outer product is
                                  1 2 3 4 5 so final result is **AVG/ι5**, that is
                                  weighted average with weights 8 4 2 1 1.

f.   4 for **AVG.MID**   **AVG/"(⊂1 2)•.MID ⊂3 4 5**
                                  ↔ AVG/"⊂3.5 4.5
                                  ↔ AVG/3.5 4.5

     7.5 for **MID.AVG**   **MID/.5×+/1 2 3 4 5**

Solutions 5d

1. a.   **∧/"CONS•.∈PROD**   b.   **∨/"CONS•.∈PROD**

To obtain the vector of vectors use

    (⊂[2]M)COMPRESS"ι¯1↑ρM

as described in Section 4.2.1.

2. a. It returns the digit sums of the first 20 positive integers.

    b. The left argument of ⊤ must be simple - the inner product means that a
second **enclose** is applied to 10 10 so that a non-simple object, viz. ⊂10 10 is
the left argument to each of the 20 separate ⊤s.

Solutions 5e

1.   [0]   Z←L(P ONLYS Q)R
        [1]   →L1 IF 1<≡Q
        [2]   →0 Z←L(P ONLY Q)R
        [3]   L1:→L2 IF~∧/0=ρQ
        [4]   →0 Z←L
        [5]   L2:Z←(L(P ONLY(↑Q))R)P ONLYS(1↓Q)R

2. a.   [0]   Z←L(P Trace)R
        [1]   ⍕'□←(Z←',LEX,' P R)''←'',LEX,''P'' R'

   b.   [0]   Z←L(P Simple)R
        [1]   →L1 IF Z←2>≡R
        [2]   →0 Z←⍕,LEX,'(P Simple)"R'
        [3]   L1:Z←⍕LEX,' P R'

3. [0]   Z←L(P Comp1 Q)R
   [1]   Z←⍕,LEX,' P Q R'

4. One solution is

[0]   Z←(P SECANT)X;T
[1]   Z←(-/(⌽X)×T)÷-/T←P¨X   ⍝ interpolate new point
[2]   Z←(⌽(×P Z)=×P¨T)/T←(↑X),Z,1↓X ⍝ select interval containing root

To solve f(x)=0 where f(x)=2-x(x-1), and using start values 1 and 7 proceed as
follows:

[0]   Z←F X
[1]   Z←2-X×X-1

        ↑F SECANT RPTUNTIL NEAR 1 7
2

Solutions 5f

1. A possible function is

[0]   Z←SHORTEN R
[1]   Z←(,⁻1↓↑¨R)' '(⁻1↑R)

2. Predicates may be defined as

[0]   Z←ISW2 R
[1]   Z←'WILLIAM'≡2⊃R

[0]   Z←SCOTCH R
[1]   Z←'MC'≡2↑(⍴R)⊃R

3.   ⊃NAMES IF~ISW2¨NAMES

4.   ∊¨⊃SHORTEN UNLESS SCOTCH¨NAMES,¨¨' '

5.   Define a function NOTSCOTCH as the negation of SCOTCH (using ~SCOTCH
won't do!) and use

        ⊃∊¨LENGTHEN UNLESS NOTSCOTCH¨NAMES,¨¨' '

[0]   Z←LENGTHEN R
[1]   Z←(⁻1↓R),⊂'MAC',2↓∊⁻1↑R

6.   (⁻1↓¨NAMES),,[⍳0]⁻1↑¨NAMES

Solutions 6a

1. Change ∆∆Isin to

```
[0]    Z←L ∆∆Isin R
[1]    →L1 IF(↑L)GT↑2⊃R
[2]    →0 Z←1+L Isin↑R
[3]    L1:Z←1+L Isin↑ΦR

       6 7 8 Isin"⊂TR1
1 3 2
       'ANN' 'DAVID'Isin"⊂TR3
2 3
```

2. The following function sequence obtains subtrees as specified:

```
[0]    Z←L SUB R
[1]    →L1 IF 0=ρR
[2]    →0 Z←L ∆SUB R
[3]    L1:Z←ι0

[0]    Z←L ∆SUB R
[1]    →L1 IF L≡2⊃R
[3]    →0 Z←L ∆∆SUB R
[2]    L1:Z←R

[0]    Z←L ∆∆SUB R
[1]    →L1 IF(↑L)GT↑2⊃R
[2]    →0 Z←L SUB↑R
[3]    L1:Z←L SUB↑ΦR

       8 SUB TR1
    7.5     8      9
       'ANN'SUB TR3
    ANN     DAVID
```

Solutions 6b

1. A function MAKENET which constructs a connectivity matrix from a vector of co-ordinate vectors is:

```
[0]    Z←L MAKENET R
[1]    Z←(LρO)∆MAKENET R

[0]    Z←L ∆MAKENET R
[1]    →L1 IF 0=ρR
[2]    →0 Z←(L ∆∆MAKENET↑R)∆MAKENET 1↓R
[3]    L1:Z←L

[0]    Z←L ∆∆MAKENET R
[1]    Z←↑L((R⌷L)←1)
```

2. a. Use say `(⊂NETL)REACH¨ι3` - answer is all nodes.

  b. Use `NETL PREC 3` - answer is node 2 only.

3.　　`T←SYMNET MST 1`
　　　`T,[.5]T⌷¨⊂SYMNET`

4. Use `MSTV` - answer is 9 by building roads 15, 54, 43 and 42.

Appendix B. Some Key Rules and Identities

Scalars and Pervasive Functions

For scalar S   S ↔ ⊂S
For pervasive F   (F R) ↔ F¨R

Indexing

Informally the shape of the result is the shape of the index and is independent of the shape of the data.

For valid I,
    for vector V   I⌷V ↔ V[I] ↔ (⊂I)⊃V
    for array A and I⌷A   (ρ,I) ↔ ρρA
    − informally the shape of the index is the rank of the array.

Operators

Operators have long left scope and short right scope, whereas functions have long right scope and short left scope.

Each

For monadic F and Z←F¨R   Z[I] ↔ ⊂F⊃R[I]
For dyadic F and Z←L F¨R   Z[I] ↔ ⊂(⊃L[I])F⊃R[I]
For scalar F, scalar S, and arrays A B C D
                                  S F¨ A B ↔ (S F A)(S F B)
                                  A B F¨ S ↔ (A F S)(B F S)
                                  A B F¨C D ↔ (A F C)(B F D)

Reduction

Reduction reduces rank, not depth.

For vector `V`	`(F/V)  ↔  V[1] F V[2] F ...`
For vector `V`	`(F/¨V)  ↔  (F/V[1])(F/V[2]) ...`
For array `A`	`(F/¨A)  ↔  ⊂F/⊃A`

Outer Product

For `Z←L∘.P R`
    for valid `I,J`   `Z[I;J]  ↔  ⊂(⊃L[I])Q⊃R[J]`

Inner Product

`L P.Q R  ↔  P/¨⊂[ρρL]L)∘.Q⊂[1]R`

Replicate

For valid `L,I,A`
    for `L/[I]A`   `(+/¯1×L)  ↔  I⊃ρA`

n-wise Reduction

For scalar `S`, vector `V`   `(S+ρS F/V)  ↔  1+ρV`

First

`(↑A)  ↔  (⊂T)⊃(T←(ρρA)ρ1)↑A`

Take/Drop

`(ρI↑A)  ↔  |I`

`(ρI↓A)  ↔  0⌈(ρA)-I`

Appendix C. List of Illustrations

Index

Reversing scans   71, 73, 147

S

Scalar extension   56, 109
Scalar functions   48, 54, 56, 57, 92, 93, 141, 152, 155
Scalarization   37, 60
Scan   71, 142, 158, 168
Scatter indexing   24, 25, 27, 66
Schoolmaster's rank   47
Scope   125, 160, 257
Secant method   183, 254
Selection   21, 111
Selective assignment   34, 39
Selective enlist   136, 196, 204
Sequences of inner products   160
Shape   9, 60, 156
Shortest path   212
Simple objects   2, 12, 239
Sorting   43
Sparse matrices   209
Spell check   94
Spirals   142
Squad indexing   24
Stem and leaf plot   21
Strand notation   6
Stretch factor   142
Structure   1, 37, 39, 107, 108, 121, 196
Structure phase   2, 115
Subtrees   200, 207
Sweeper   145

T

Take   23, 103, 258
Tied ranks   46
Titling   63
Tracing function execution   122, 132
Tree operations   201
Trees   195, 196
Type   39, 107

U

Undirected networks   215
Universal quantifier   153
User-defined operators   121

V

Vector assignment   34
Vector notation   5, 23

W

Weighted moving average   58, 232
Without   21, 32, 50, 210
Word search   94